U.S. Fire Administration
Mission Statement

As an entity of the Department of Homeland Security, the mission of the USFA is to reduce life and economic losses due to fire and related emergencies, through leadership, advocacy, coordination, and support. We serve the Nation independently, in coordination with other Federal agencies, and in partnership with fire protection and emergency service communities. With a commitment to excellence, we provide public education, training, technology, and data initiatives.

Homeland
Security

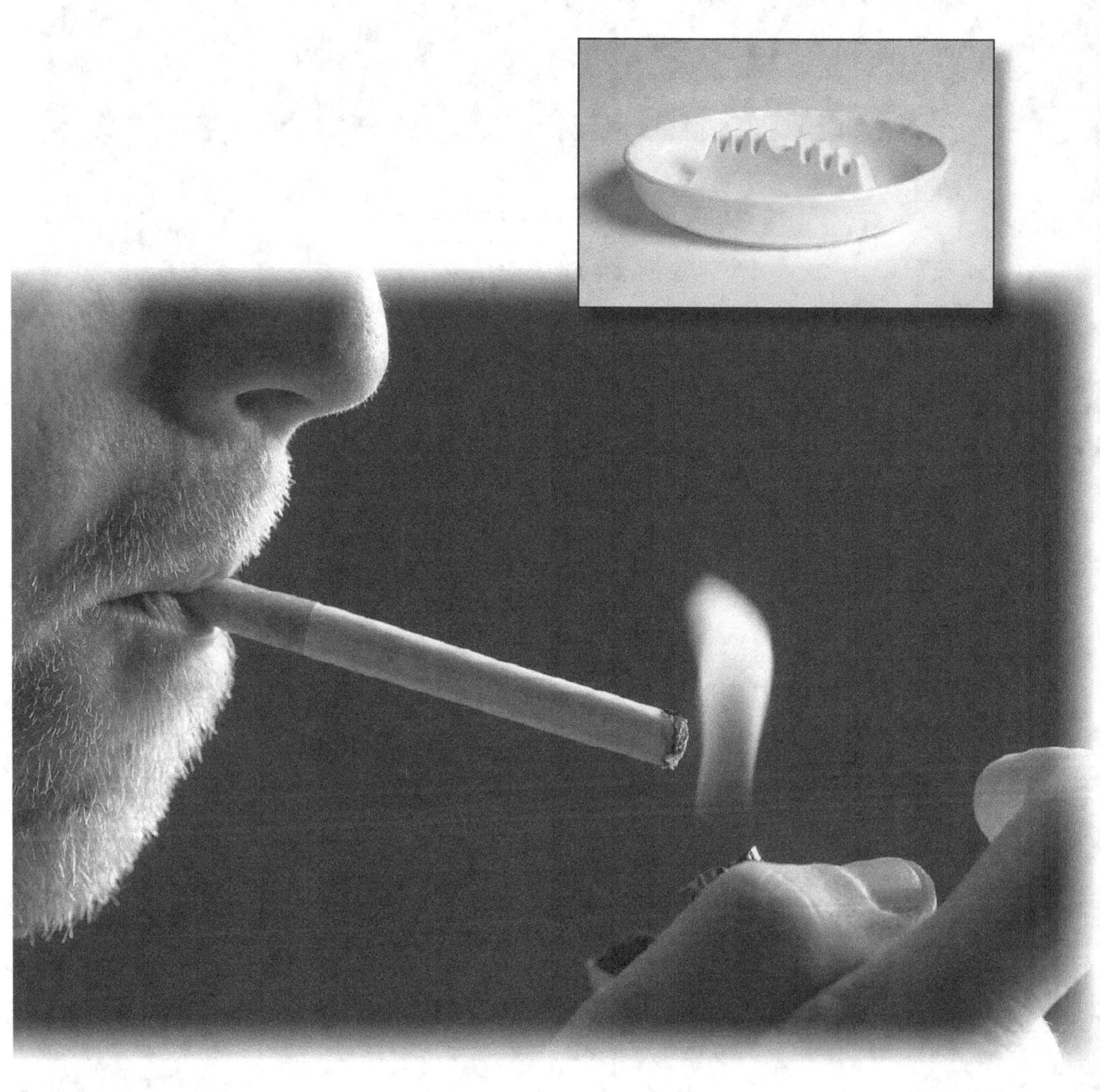

Behavioral Mitigation of Smoking Fires Through Strategies Based on Statistical Analysis

Behavioral Mitigation of Smoking Fires Through Strategies Based on Statistical Analysis

Final Project Report for EME-2003-CA-0310

John R. Hall, Jr.
Marty Ahrens
Kimberly Rohr
Sharon Gamache
Judy Comoletti
National Fire Protection Association

February 2006

Executive Summary

Fires started by lighted tobacco products, principally cigarettes, constitute the leading cause of residential fire deaths. The U.S. Fire Administration (USFA) has partnered with the National Fire Protection Association (NFPA) "to research what types of behaviors cause smoking fire fatalities and develop sound recommendations for behavioral mitigation strategies to reduce smoking fire fatalities in the United States...."

The scope of the study included all lighted tobacco products, but cigarettes account for nearly all consumption and fires. Lighting implements such as matches and lighters were not included. Most fires involving these objects occur during incendiarism or fireplay, and only a small fraction occur in the process of lighting cigarettes.

Smoking-material home fire deaths are almost three times as likely as home fire deaths caused by other means to involve a victim who was close to the ignition (29 percent versus 11 percent in 1994 to 1998). Fatal victims of smoking-material fires are, therefore, less likely than fatal victims of other kinds of fires to be saved by strategies and technologies that react after ignition, i.e., fire protection devices. For many, if not most, of these victims, there is no substitute for prevention.

Smokers are more likely to have unrelated impairments, limitations, disabilities, or other characteristics that can interfere with their response to fire. This can mean a more serious injury for a defined level of exposure to fire.

The majority of smoking-material home structure fires and more than two-thirds of associated deaths involve trash, mattresses, bedding, or upholstered furniture as the first ignited item. Both mattresses and upholstered furniture have been the subject of decades-long requirements, industry-based or government-based, respectively, to reduce ignitability by cigarettes. The long-term impact of these programs can be seen in the rising percentage of fatal home structure smoking-material fires that begin with ignition of something other than upholstered furniture, mattresses, or bedding. That percentage

was 15 percent in 1980 to 1982, 20 percent in 1990 to 1992, and 29 percent in 2000 to 2002.

The characteristics of an effective ashtray have been described by many different terms in existing educational materials, but some of those terms (e.g., "large") were judged to be both vague and potentially inadequate. An ashtray is intended to provide a safe repository for ashes while a cigarette is being smoked and a safe temporary repository for ashes and butts after a cigarette has been smoked. This will happen if the ashtray minimizes

- the likelihood of a lit cigarette falling out of the ashtray (depth was deemed the most important feature);

- the likelihood of the ashtray itself overturning and spilling ashes, embers, and butts onto potential combustibles ("sturdy" was deemed the best established term for what is needed in such an ashtray); and

- the likelihood of a hostile fire if ashes, embers, or butts fall outside the protected confines of the ashtray (a sturdy, hard-to-ignite surface for the ashtray was deemed the best way to describe what was needed).

Nearly half of all smoking-material home structure fires and roughly three-fourths of associated deaths involve fires that begin in the bedroom, living room, family room, or den. Most fatal victims were asleep when fatally injured but most fatal smoking-material home structure fires did not begin in the bedroom.

Available data do not permit calculation of the risk of fatal cigarette fires relative to time spent smoking, distinguishing different rooms of a home. There is some indirect evidence of the relative safety of smoking outdoors. In 1994 to 1998, in the winter months of December through February, the rate of deaths per 100 reported smoking-material home structure fires was much higher--4.9 in December through February and 2.8 in the other months, or 75 percent higher in winter. Winter is when going outdoors to smoke is most difficult and, therefore, least likely to happen.

There is some evidence that a message regarding where to smoke will, in many cases, be seen as reinforcing established household rules, rather than reinventing behavior.

Two major Federal-government studies of the potential risk reduction from a reduced ignition-strength cigarette reached the following conclusions:

- A reduced ignition-strength cigarette is technically feasible.

- A standard test of cigarette ignition strength is technically feasible.

The smoker whose smoking materials ignited the fire is the only person present in just over half of fatal cigarette fires. Even for these 54 percent of cases, smokers may not live alone and may be influenced by others in the behaviors that led to ignition. In the 46 percent of cases where someone else is present, it was not known whether those others had characteristics that would affect their ability to exert such influence effectively.

One fatal victim in four (24 percent) is **not** the smoker whose cigarette started the fire. Therefore, if others are present, they have both a direct and an indirect stake in taking action to prevent hostile fires from taking place.

The relationships of these victims to the smokers is useful to know because it may bear on the willingness and ability of these others to serve as "watchers" for the smokers, as well as the willingness of the smokers to accept help or advice from these others.

Of the fatal victims who were not the smokers whose smoking materials ignited the fires:

- Thirty-four percent were children of the smokers (that is, the smokers were the parents of the victims, but some of these victims were themselves adults).

- Twenty-five percent were neighbors (often from other apartment units in the same building) or friends of the smokers.

- Fourteen percent were spouses or partners of the smokers.

- Thirteen percent were parents of the smokers.

- Fourteen percent had other relationships (e.g., sibling, niece or nephew, uncle or aunt, roommate, passerby).

The project recommends the use of four general messages, two more specific messages for particular audiences, and a seventh message to be used when space or time permit. The four general messages are

If you smoke, smoke outside.

Wherever you smoke, use deep, sturdy ashtrays. Ashtrays should be set on something sturdy and hard to ignite, like an end table.

Before you throw out butts and ashes, make sure they are out, and dowsing in water or sand is the best way to do that.

Check under furniture cushions and in other places people smoke for cigarette butts that may have fallen out of sight.

The two specific messages:

Smoking should not be allowed in a home where oxygen is used.

If you smoke, choose fire-safe cigarettes. They are less likely to cause fires.

And the seventh message:

To prevent a deadly cigarette fire, you have to be alert. You won't be if you are sleepy, have been drinking, or have taken medicine or other drugs.

These messages have been applied to existing USFA educational materials (see Appendix E) and are being adopted into NFPA educational messages as they come up for routine revision.

The project has developed two PowerPoint® presentations--one for educators and one for smokers and others whose behavior we seek to change--to implement the recommended educational messages and provide photographs for added clarity. (See Attachments I and II.)

Table of Contents

Introduction

Fires started by lighted tobacco products, principally cigarettes, constitute the leading cause of residential fire deaths. The U.S. Fire Administration (USFA) has partnered with the National Fire Protection Association (NFPA) "to research what types of behaviors cause smoking fire fatalities and develop sound recommendations for behavioral mitigation strategies to reduce smoking fire fatalities in the United States...."

The scope of the study included all lighted tobacco products, but cigarettes account for nearly all consumption and fires. Lighting implements such as matches and lighters were not included. Most fires involving these objects occur during incendiarism or fireplay.

An extensive literature review on behaviors related to smoking, or to fires or fatalities due to smoking-material fires was conducted to provide the broadest possible fact base for recommendations. In addition, data were collected from:

- analysis of the 1980 to 2001 U.S. smoking-material fire problem, using The National Fire Incident Reporting System (NFIRS) national estimates;

- analysis of several hundred 1997 to 1998 fatal smoking-material fires, not necessarily representative but documented in greater detail in NFPA's major fires database called the Fire Incident Data Organization (FIDO);

- analysis of other risk factors correlated with smoking, based on the U.S. Centers for Disease Control and Prevention (CDC) Behavioral Risk Factor database for 2002.

The two detailed fire incident databases used in these analyses are described below and in slightly more detail in Appendix A:

NFIRS: This USFA database is the most representative national fire database, providing detailed information on individual fires and casualties. Nearly all national estimates of specific aspects of the U.S. fire problem begin with NFIRS. About a third of U.S. fire departments--working through their respective States--participate in NFIRS, which receives reports on an estimated one-third to one-half of each year's fires. The NFPA and most other users of NFIRS combine it with the NFPA survey to produce the best "national estimates" of the specific characteristics of the Nation's fire problem.

Some important coding information, such as the victim location code "intimate with ignition," are not coding options in NFIRS Version 5.0, which applies to 1999 and later data. Therefore, pattern analysis in this report is done using 1994 to 1998 data throughout.

FIDO: Many questions of technical interest require a level of detail beyond that available through NFIRS. For these, the best approach often is to use exploratory data with sufficient validated detail to be useful, even if it may not be representative of overall U.S. fire experience. The largest such database, excluding proprietary insurance-industry databases, is NFPA's FIDO. In 1997 to 1998, FIDO was set up to include all fatal fires. For this project, data were extracted from 300 qualified 1997 to 1998 incidents, with records on 477 individuals, including 389 deaths.

The development of recommended standard messages and behaviors was accomplished by the NFPA Educational Messaging Advisory Committee (EMAC), a diverse group of educational experts recently formed to advise the NFPA on educational messaging across the full range of fire and life safety education programs. They were provided with a briefing on the results of this project's research.

The project also incorporated concepts from a Health Belief model widely used in public health research and initiatives. The model is built around six generic questions:

1

1. How likely does the individual consider the kind of harm targeted by the strategy to be?

2. How serious does the individual consider the kind of harm targeted by the strategy to be?

3. How much benefit does the individual believe he/she would derive from a change in the targeted behavior?

4. What adverse side effects of a change in behavior are perceived as barriers by the target audience?

5. What cues to action are part of the strategy?

6. How confident is the target audience that behavior can be changed?

Finally, near the end of the project, results were provided from a parallel effort by Hager Sharp, the public relations firm, which included a focus group review of candidate messages related to smoker behavior. Hager Sharp obtained information no other source had provided on the likely receptiveness of smokers to the tested messages and other similar messages.

A 2005 USFA statistical report--*Residential Smoking Fires and Casualties*--is complete and available at http://www.usfa.fema.gov/downloads/pdf/tfrs/v5i5.pdf. It provides additional statistical material on the smoking fire problem.

The U.S. Smoking-Material Fire Problem

Size and Trends of the Fire Problem

Smoking-material structure fires and associated civilian deaths have declined sharply since 1980 (Figure 1), but smoking materials remain the leading cause of structure fire deaths in the United States. Homes account for more than two-thirds of smoking-material structure fires and more than 90 percent of smoking-material structure fire deaths. Little is lost by focusing on the home portion of the problem. (Any unreferenced statistics are national estimates from NFIRS and the NFPA survey. See Appendix B for more details.)

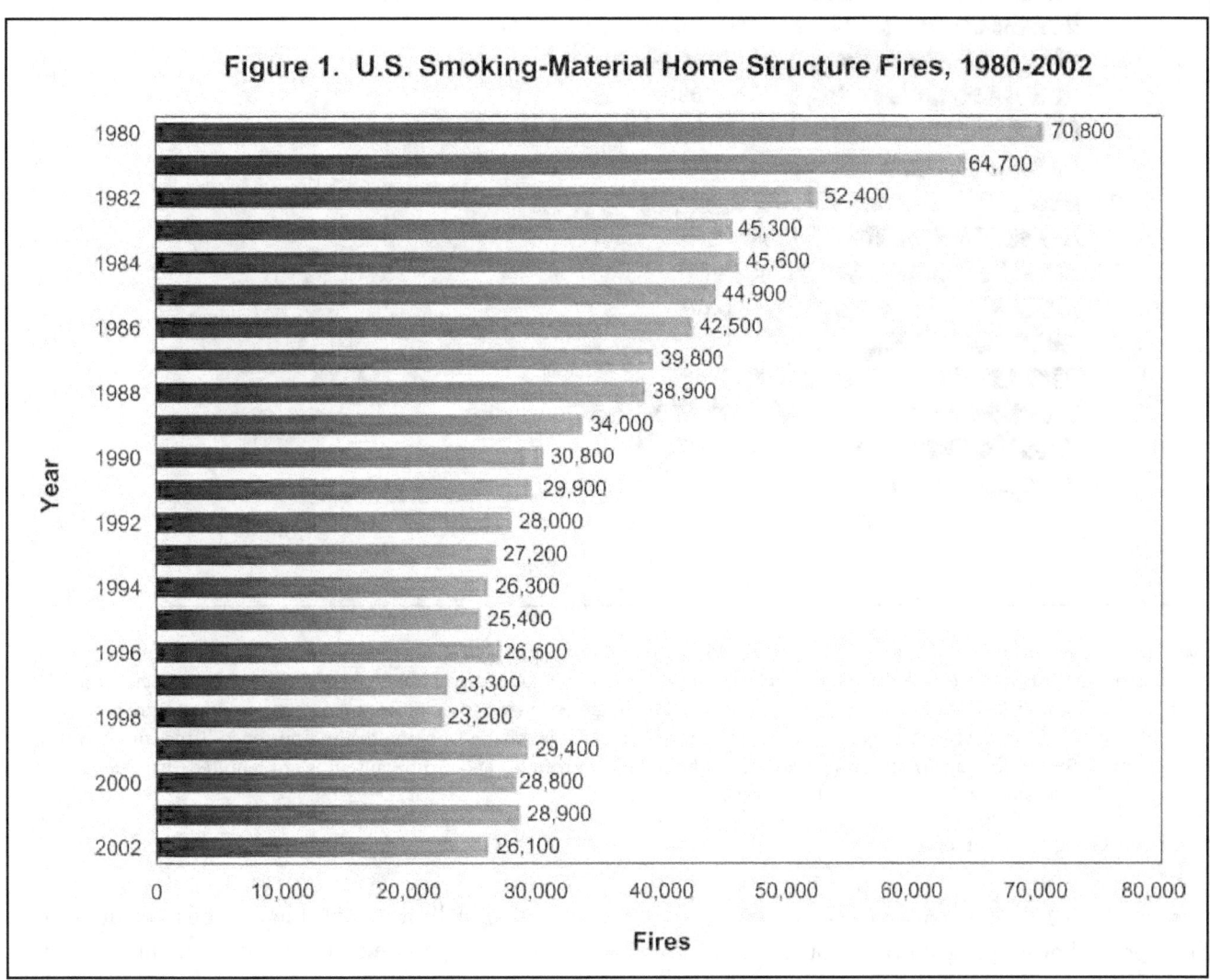

Figure 1. U.S. Smoking-Material Home Structure Fires, 1980-2002

Note: These are national estimates of fires reported to U.S. municipal fire departments. Fires reported only to Federal or State agencies or industrial fire brigades are excluded. National estimates are projections. Casualty and loss projections can be influenced heavily by the inclusion or exclusion of one unusually serious fire. Fires are estimated to the nearest hundred. Figures reflect a proportional share of structure fires in which the heat source was unknown, and fires involving smoking materials or open flames of unknown type. From 1999 on, confined trash-receptacle fires--with the heat source not reported because it is not required--have been proportionally allocated based on 1994 to 1998 heat-source patterns.

Source: NFIRS and NFPA survey

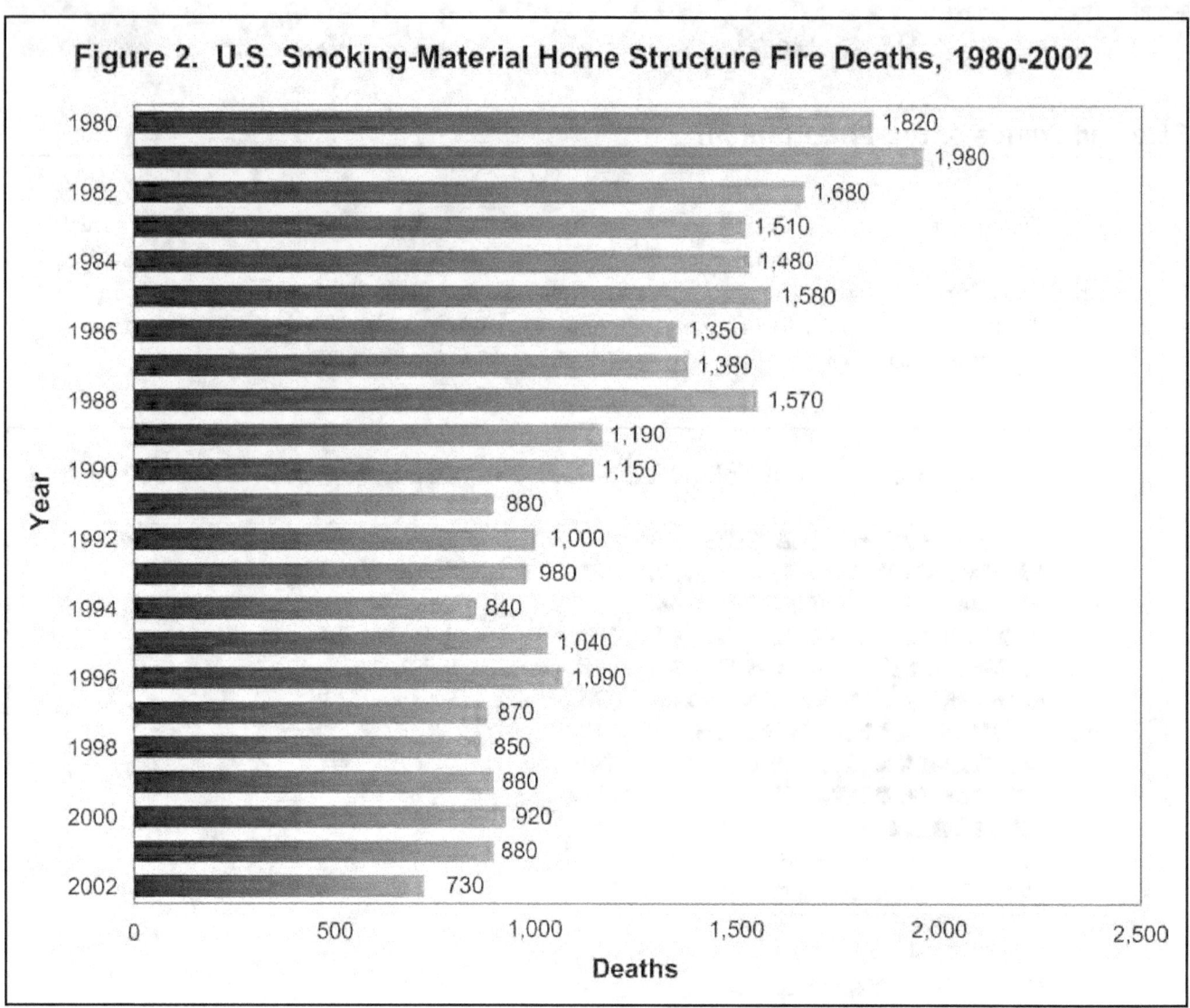

Figure 2. U.S. Smoking-Material Home Structure Fire Deaths, 1980-2002

Note: These are national estimates of fires reported to U.S. municipal fire departments. Fires reported only to Federal or State agencies or industrial fire brigades are excluded. National estimates are projections. Casualty and loss projections can be influenced heavily by the inclusion or exclusion of one unusually serious fire. Deaths are estimated to the nearest ten. Figures reflect a proportional share of structure fires in which the heat source was unknown, and fires involving smoking materials or open flames of unknown type. From 1999 on, confined trash-receptacle--fires with the heat source not reported because it is not required--have been proportionally allocated based on 1994 to 1998 heat-source patterns.

Source: NFIRS and NFPA survey.

The number of smoking-material home structure fires declined by 63 percent from 1980 to 2002 (Figure 2). The number of smoking-material home structure fire deaths declined by 60 percent in the same period. (See Tables B-1 and B-2 in Appendix B for deaths and injuries by age and sex.)

More than half of the decline may be attributed to declines in cigarette consumption. The number of cigarettes consumed fell by 28 percent from 1987 to 2002.* Nearly all smoking-material fires and losses involve cigarettes.

---

* For 1987 and 1988 data, John C. Maxwell, Jr., *The Maxwell Consumer Report: 1988 Year-End Sales Estimates for the Cigarette Industry,* Richmond, Virginia: Wheat First Securities, January 27, 1989; for 1994 to present, Tom Capehart, *Tobacco Outlook,* U.S. Department of Agriculture, www.ers.usda.gov, October 6, 2003; and for 1989 to 1993, earlier reports in the *Tobacco Situation and Outlook Report* series.

There Is No Substitute for Prevention When a Victim is "Intimate With Ignition"

Smoke alarms, sprinklers, and compartmentation barriers all require time after ignition to be effective. For a victim recorded as "intimate with ignition," the fire begins so close to him or her that it is very difficult to survive long enough for active or passive fire protection to save him or her.

From 1994 through 1998, smoking-material home fire deaths were almost three times as likely as other-cause home fire deaths to involve a victim intimate

with ignition (29 percent versus 11 percent). (See Table B-3 in Appendix B.)

Being intimate with ignition gives a person minimal time to react effectively to a threatening fire, and various conditions reduce a person's ability to use whatever time he or she has. The most common fatal-victim characteristic with this effect is sleeping (Figure 3). Smoking ranks first among the 12 leading causes of fire in home fire deaths (1994 to 1998) with 23 percent. If you specify that the victim was sleeping before injury, this rises to 27 percent. (See Table B-4 in Appendix B.)

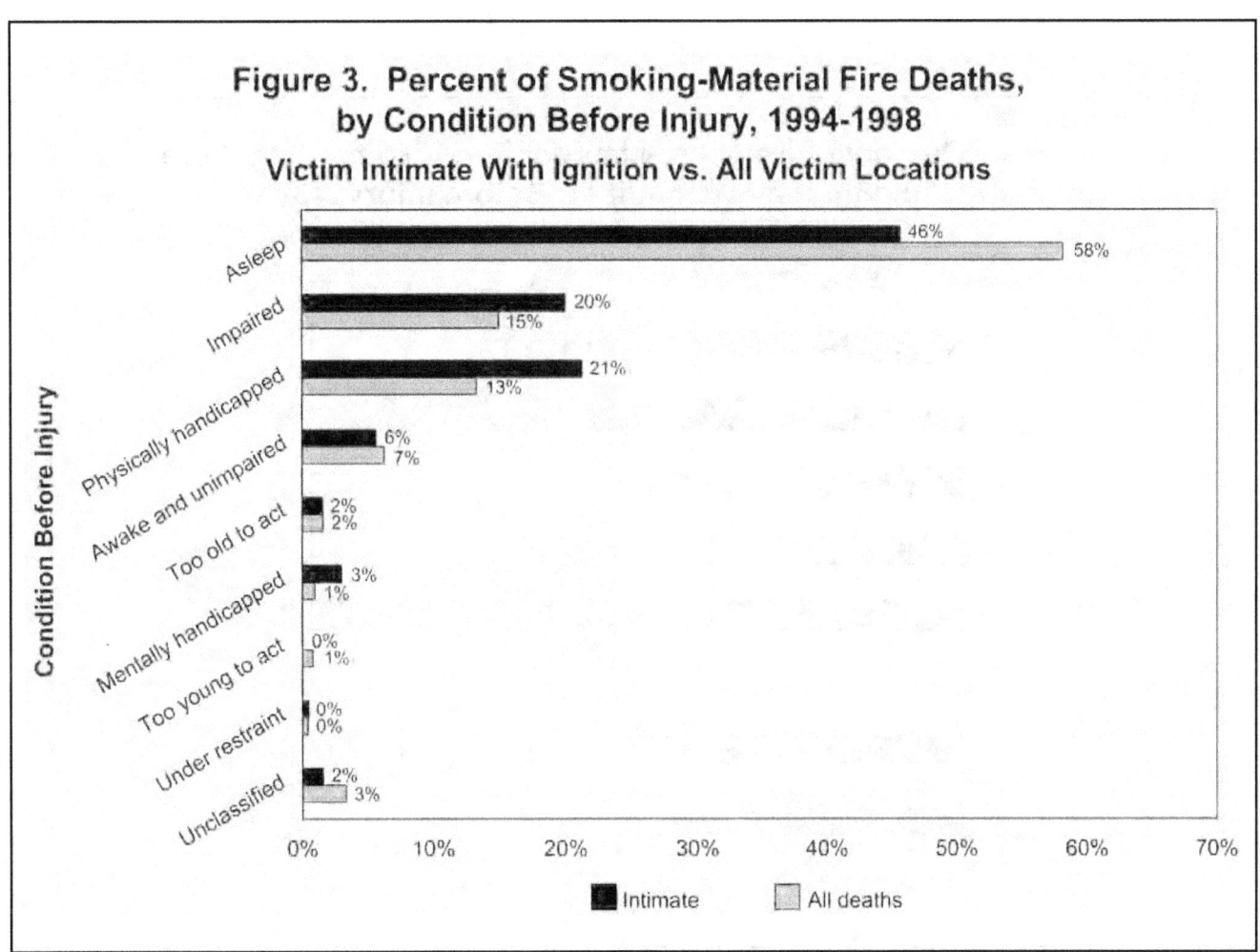

Figure 3. Percent of Smoking-Material Fire Deaths, by Condition Before Injury, 1994-1998. Victim Intimate With Ignition vs. All Victim Locations

Note: These are national estimates of fires reported to U.S. municipal fire departments. Fires reported only to Federal or State agencies or industrial fire brigades are excluded. National estimates are projections. Casualty and loss projections can be influenced heavily by the inclusion or exclusion of one unusually serious fire. Figures reflect a proportional share of home smoking-material structure fire deaths in which the victim's condition before injury was unknown.

Source: NFIRS and NFPA survey.

The heightened risk associated with being intimate with ignition can be compounded by certain conditions, activities, or other characteristics. For example, of the smoking-material home-fire fatal victims who were intimate with ignition, nearly half were asleep before injury (45 percent) and most of the rest were physically or mentally handicapped or impaired by alcohol or other drugs (44 percent). (See Table B-5 in Appendix B.) Being asleep compounds the risk from being close to the fire, and having specific age-related limitations, disabilities, or impairments compounds that risk even more. For all smoking-material home-fire fatal victims, roughly six-tenths (58 percent) were asleep before injury and

roughly three-tenths (29 percent) were physically or mentally handicapped or impaired by alcohol or other drugs. (See Table B-6 in Appendix B.) Both percents are higher than for other-cause fire deaths.

Put another way, victims of fatal smoking-material fires are less likely to have been intimate with ignition if they were awake and unimpaired (26 percent) or asleep (23 percent) than if they had a disability (50 percent for physical handicap, 57 percent for mental handicap), were restrained (44 percent), were impaired by alcohol or other drugs (40 percent), or had the physical and mental limitations associated with old age (34 percent) (Figure 4).

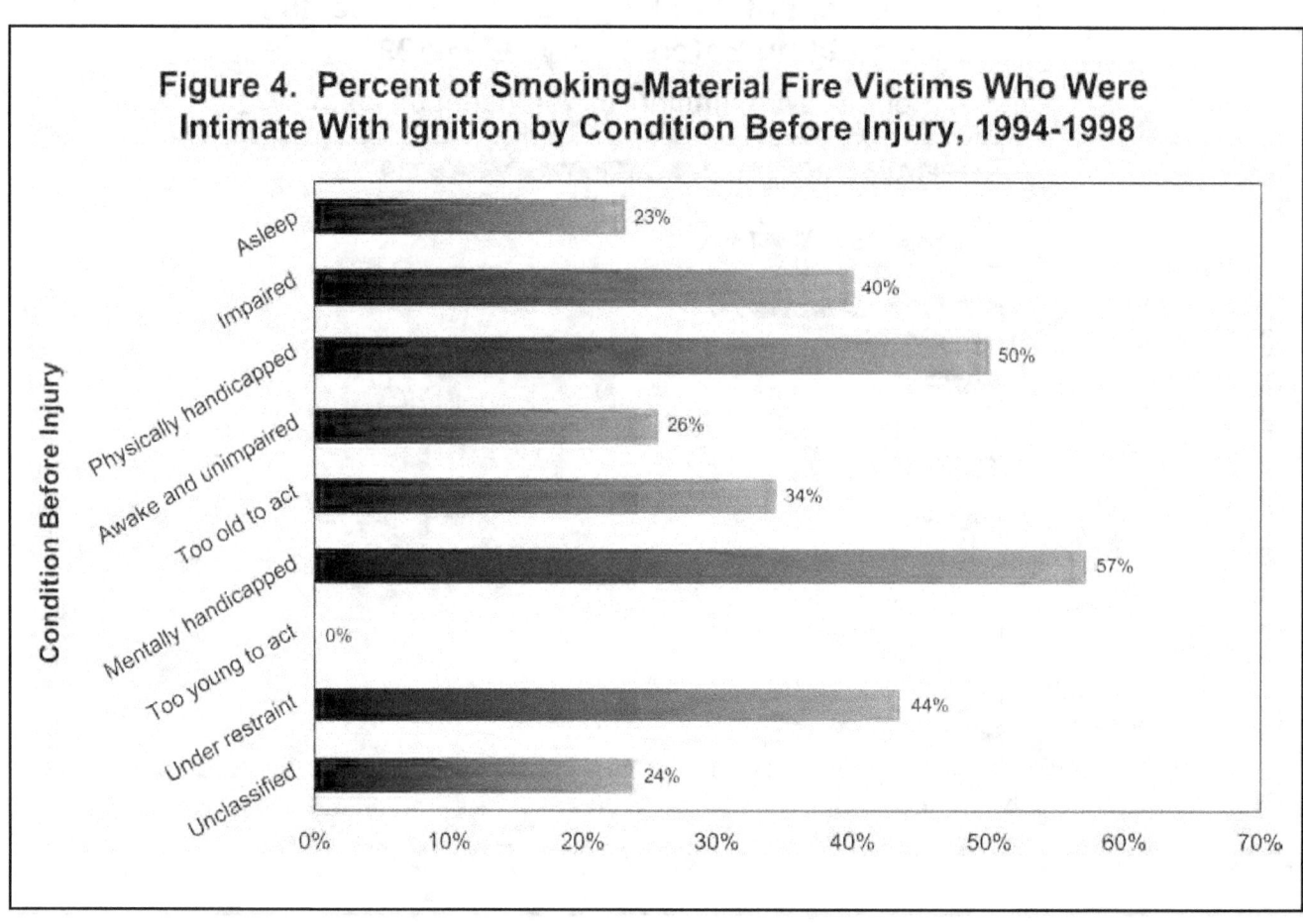

Figure 4. Percent of Smoking-Material Fire Victims Who Were Intimate With Ignition by Condition Before Injury, 1994-1998

Note: These are national estimates of fires reported to U.S. municipal fire departments. Fires reported only to Federal or State agencies or industrial fire brigades are excluded. National estimates are projections. Casualty and loss projections can be influenced heavily by the inclusion or exclusion of one unusually serious fire. Figures reflect a proportional share of home smoking-material structure fire deaths, for each victim condition before injury, in which the victim location was unknown.

Source: NFIRS and NFPA survey.

The link between smoking and sleeping is even stronger if the fire started because someone fell asleep when he or she should have been supervising a heat source. When ignition factor was coded as falling asleep, three-fourths of the fatal home fires in 1994 to 1998 were smoking-material fires. (See Table B-4 in Appendix B.) (The majority of the rest were cooking fires.) And when falling asleep is the reason a fire started, it is not surprising if the smoker is intimate with that ignition and at very high risk of fatal injury. For smoking-material home fires where falling asleep was coded as the ignition factor, 41 percent of fatal victims were intimate with ignition, compared to 29 percent of fatal victims of smoking-material fires generally and 23 percent of all sleeping fatal victims of smoking-material fires. (See Figure B-1 in Appendix B.)

As discussed, a fire that begins very close to a person requires very little time to grow large enough to cause a fatal injury, unless the ignited materials are designed to burn slowly. There is no compartmentation separating person from fire, so compartmentation does not help. Even if a smoke alarm should activate before a person is fatally injured, the statistics just presented show 44 percent of fatal intimate-with-ignition victims have some serious condition--a disability or impairment--that would make a successful escape attempt unlikely in the very short time available. In addition, in order to activate, a fire sprinkler requires more severe fire conditions at the ceiling than a smoke alarm does. As a result, even though the sprinkler will provide more direct and immediate protection, not requiring any action on the occupant's part to save himself or herself, it is doubtful that an intimate-with-ignition victim will be saved by a sprinkler.

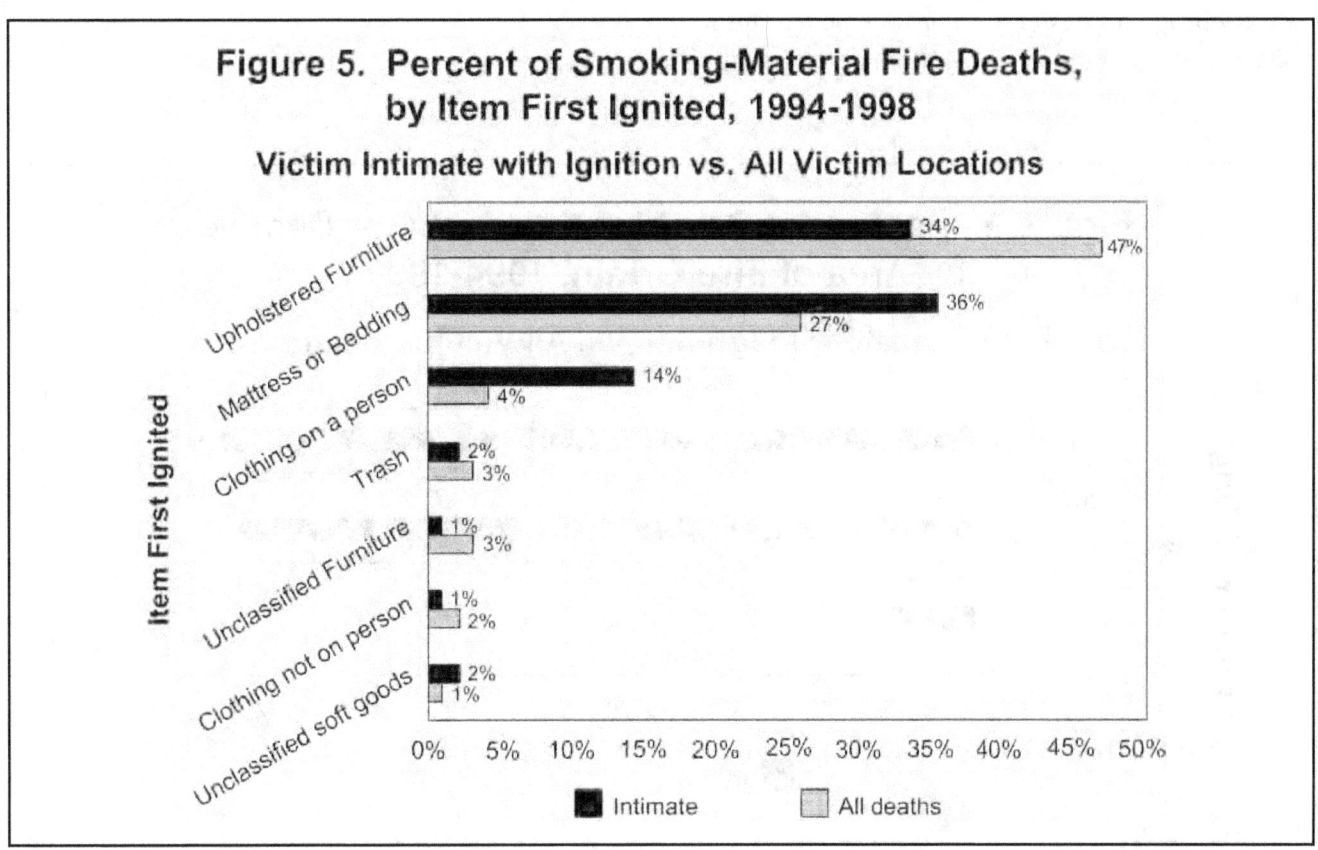

Figure 5. Percent of Smoking-Material Fire Deaths, by Item First Ignited, 1994-1998

Victim Intimate with Ignition vs. All Victim Locations

Note: These are national estimates of fires reported to U.S. municipal fire departments. Fires reported only to Federal or State agencies or industrial fire brigades are excluded. National estimates are projections. Casualty and loss projections can be influenced heavily by the inclusion or exclusion of one unusually serious fire. Figures reflect a proportional share of home smoking-material structure fires with item first ignited unknown.

Source: NFIRS and NFPA survey.

However, one possible exception to the seeming inevitability of fatal injury when fire begins close to the person is when the fire involves materials that are designed to burn slowly. These materials, such as slower-burning upholstered furniture or mattresses, allow more time for some type of protective action to be taken, despite the close proximity of a fire to a potential victim.

Most fatal smoking-material home fires begin with ignition of upholstered furniture, a mattress or bedding, or clothing (Figure 5). These items account for 80 percent of all smoking-material home fire deaths and 85 percent of such deaths when the victim is intimate with ignition. (See Table B-7 in Appendix B.) Within these groups, the relative importance of clothing and mattress or bedding is greater for the intimate-with-ignition victims than for all victims.

Since the 1960's, there have been regulations or industry programs designed to provide better ignition resistance to cigarettes and slower fire development for upholstered furniture and mattresses. As one might expect, the share of smoking-material home fire deaths involving initial ignition of upholstered furniture, mattress, or bedding has declined from five out of six in the early 1980's to three out of four in the late 1990's. (See Table B-8 in Appendix B.) These initiatives are part of the reason for the decline in overall smoking-material home fire deaths.

The largest share of fatal home smoking-material fire victims intimate with ignition are in the living room, family room, or den (44 percent) compared with the bedroom (42 percent), although the bedroom share is larger than for all home smoking-material fire deaths (42 percent versus 36 percent). Note also that the kitchen share is larger for intimate-with-ignition fire deaths, suggesting that cigarettes are being dropped on the victim's clothing outside the bedroom (Figure 6).

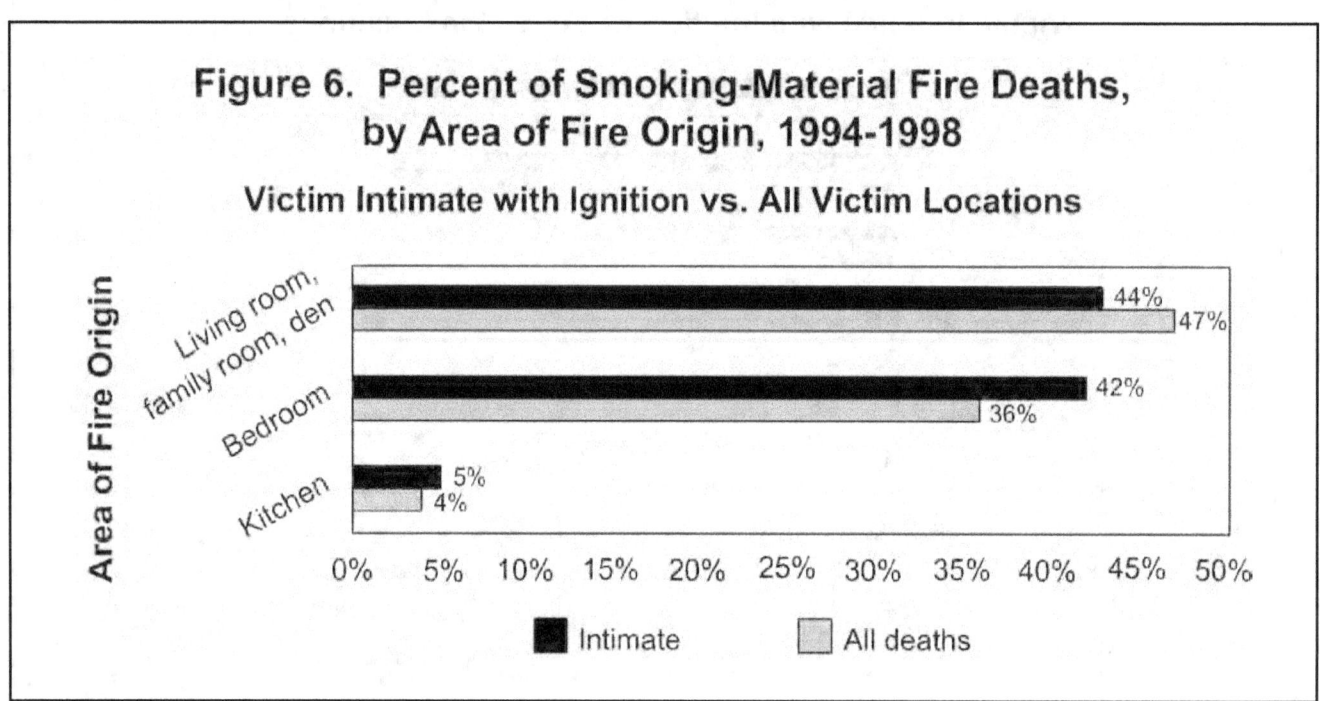

Figure 6. Percent of Smoking-Material Fire Deaths, by Area of Fire Origin, 1994-1998

Victim Intimate with Ignition vs. All Victim Locations

Note: These are national estimates of fires reported to U.S. municipal fire departments. Fires reported only to Federal or State agencies or industrial fire brigades are excluded. National estimates are projections. Casualty and loss projections can be influenced heavily by the inclusion or exclusion of one unusually serious fire. Figures reflect a proportional share of home smoking-material structure fires in which area of fire origin unknown.

Source: NFIRS and NFPA survey.

From 1994 to 1998, fatal victims of smoking-related home fires whose activity when injured was **sleeping** were roughly 10 times as likely as sleeping victims of fires due to other causes to be intimate with ignition (30 percent versus 3 percent). (See Table B-9 in Appendix B.) (The 30 percent for victims whose **activity** when injured was sleeping differs from the 23 percent cited earlier for victims whose **condition** before injury was sleeping.)

All of these findings support the general conclusion that fatal victims of smoking-material fires are less likely than fatal victims of other kinds of fires to be saved by strategies and technologies that react after ignition, i.e., fire protection provisions. For many if not most of these victims, there is no substitute for prevention.

Correlated Characteristics of Smokers Make Effective Response to Fire Less Likely and Serious Consequences From Fire More Likely

Smokers in general, smokers who have fires, and smokers who are fatally injured in fires, all are more likely to have unrelated impairments, limitations, disabilities, or other characteristics that can interfere with their response to fire or can result in a more serious injury for a defined level of exposure to fire effects. (See Tables B-10 and B-11 in Appendix B for more statistics on these characteristics.)

These characteristics do not point to particular smoker behaviors that should be modified on the grounds of reducing risk of fire death. However, they all point to behaviors that, if modified, would result in less risk of harm. In addition, all of these characteristics increase the need for some more effective means of avoiding cigarette fires and associated casualties.

Impairment by Alcohol or Other Drugs

- According to NFIRS national estimates for 1994 to 1998, fatal victims of home smoking-material fires are more likely than victims of other fatal home fires to be impaired by alcohol or other drugs (15 percent versus 7 percent). (See Table B-6 in Appendix B.) It is known that NFIRS tends to under-report alcohol and drug impairment.

- Three special studies, designed to use blood-alcohol tests to provide a more complete and accurate estimate of alcohol impairment in fatal fire victims, consistently found much higher rates of alcohol involvement in victims of fires in general and in smoking-related fires in particular. They cited alcohol as a factor in smoking-related fire deaths in almost half the deaths in London, UK, in 1996 to 2000; almost half the deaths in Tallahassee, FL, in 1983 to 1994; and 62 percent of the deaths in Minnesota in 1996 to 2002.[1,2,3]

- Fatal victims who were also the smokers whose cigarettes started the fires are much more likely to have been impaired than fatal victims who were not those smokers. According to FIDO, 47 percent of the deaths of smokers whose smoking material ignited the fire involved alcohol or other drug use as a factor. Specifically, those deaths consisted of 41 percent alcohol only, 3 percent other drugs only, and 3 percent alcohol and other drugs. Alcohol or other drug use was cited for 10 percent of the nonsmoker deaths, and all were alcohol-only. (Tables B-12 to B-15 in Appendix B contain a listing of tallies of the FIDO cases by combinations of characteristics. Table B-14 has the particular listings used to calculate these percentages.)

- While alcohol and other drugs can have a strong soporific effect (i.e., causing drowsiness), many fatal victims of smoking-material fires appear to have been drowsy even apart from the effects of alcohol. According to FIDO (see Table B-14 in Appendix B), 26 percent of smoker deaths involved alcohol and evidence of sleepiness independent of alcohol effects, as did 6 percent of nonsmoker deaths.

- According to the CDC, smokers defined as those who have smoked at least 100 cigarettes in their lifetimes were more likely than nonsmokers to have consumed five or more alcoholic drinks at one occasion (29 percent versus 19 percent). These smokers also averaged one more drink per occasion than nonsmokers (3.72 versus 2.78 drinks per occasion). These statistics indicate that smokers are more likely than nonsmokers to have alcohol-impaired judgment and ability when they drink.

- According to the CDC, smokers defined as everyday or someday smokers were slightly **less** likely than nonsmokers to have had more than five drinks on an occasion (40 percent versus 41 percent), but they still averaged 1/3 to 1/2 more drinks per occasion than nonsmokers (4.76 versus 4.34 drinks per occasion).

Physical Handicaps or Limitations

- In 1994 to 1998, according to NFIRS national estimates, fatal victims of home smoking-material fires were more likely than victims of other fatal home fires to be physically handicapped (13 percent versus 7 percent), but less likely to have the physical and mental limitations associated with old age (2 percent versus 3 percent). (See Table B-6 in Appendix B.) Until recently, however, fire departments could not report multiple conditions of victims to NFIRS. As a result, the extent of these disabilities and age-related limitations most likely are understated.

- According to FIDO (see Table B-14 in Appendix B), 30 percent of smoker deaths involved physical disabilities not related to age or physical limitations related to age. Specifically, these deaths consisted of 15 percent physical disabilities only and 15 percent physical limitations only. A similar 28 percent of nonsmoker deaths involved these disabilities and limitations. Specifically, nonsmoker deaths consisted of 5 percent physical disabilities only, 22 percent physical limitations due to age only, and 1 percent both physical disabilities and age-related physical limitations.

- A 2001 study of teenagers found 31 percent of those who were mobility-impaired smoked compared to 20 percent of a comparison group without handicaps or limitations.[4]

- According to the CDC, smokers who have smoked at least 100 cigarettes in their lifetimes were more likely than nonsmokers to have the following potentially mobility-related physical handicaps or limitations:

Handicap/Limitation	Smoker	Nonsmoker
Ever told you have arthritis	35.0%	27.9%
Activity limitation due to physical, mental, or emotional problems	21.1%	14.3%
Limitations due to arthritis or other joint symptoms	30.1%	25.7%
Never exercised in the past month	27.7%	23.3%
Physical health not good for at least one day in the past 30 days	36.1%	32.0%
Health problems that require the use of special equipment	7.5%	5.7%

- According to the CDC, smokers defined as everyday or someday smokers were more likely than nonsmokers to have the following physical handicaps or limitations:

Handicap/Limitation	Smoker	Nonsmoker
Ever told you have arthritis	31.1%	26.7%
Activity limitation due to physical, mental, or emotional problems	21.3%	17.6%
Limitations due to arthritis or other joint symptoms	31.7%	29.9%
Never exercised in the past month	34.5%	22.7%
Physical health not good for at least one day in the past 30 days	38.8%	38.2%
Health problems that require the use of special equipment	6.1%	5.8%

Mental or Emotional Handicaps or Limitations

- In 1994 to 1998, according to NFIRS national estimates, fatal victims of home smoking-material fires were **less** likely than victims of other home fatal fires to be mentally handicapped (1 percent versus 2 percent) or to have the physical and mental limitations associated with old age (2 percent versus 3 percent). (See Table B-6 in Appendix B.) Until recently, however, fire departments could not report multiple conditions of victims to NFIRS. As a result, the extent of disabilities and age-related limitations most likely are understated.

- According to FIDO (see Table B-14 in Appendix B), 3 percent of smoker deaths involved mental handicaps not related to age or mental limitations related to age. Specifically, these deaths consisted of 2 percent mental disabilities only and 1 percent mental limitations only. A similar 4 percent of nonsmoker deaths involved these handicaps or limitations. Specifically, these nonsmoker deaths consisted of 1 percent mental disabilities only and 3 percent mental limitations due to age only.

- A 2001 study of teenagers found 33 percent of those who were emotionally disabled and 27 percent of those who were learning-disabled smoked compared to 20 percent of a comparison group without handicaps or limitations.[4]

- According to the CDC, smokers defined as those who have smoked at least 100 cigarettes in their lifetimes were more likely than nonsmokers to report that their mental health had not been good for at least 1 day in the past 30 days (33.5 percent versus 29.3 percent).

- According to the CDC, smokers defined as everyday or someday smokers were more likely than nonsmokers to report that their mental health had not been good for at least 1 day in the past 30 days (41.1 percent versus 39.8 percent).

Generally Poor Pre-Existing Health

- According to the CDC, smokers defined as those who have smoked at least 100 cigarettes in their lifetimes were more likely than nonsmokers to have the following physical conditions that could make them more susceptible to harm from a defined exposure to fire effects:

Physical Condition	Smoker	Nonsmoker
High blood cholesterol	37.1%	31.0%
Physical health not good for at least 1 day in past 30	36.1%	32.0%
High blood pressure	30.0%	26.5%
Asthma	13.0%	11.1%
Diabetes	8.3%	6.9%

- According to the CDC, smokers defined as everyday or someday smokers were more likely than nonsmokers to have the following physical conditions that could make them more susceptible to harm from a defined exposure to fire effects:

Physical Condition	Smoker	Nonsmoker
High blood cholesterol	32.1%	27.6%
Physical health not good for at least 1 day in past 30	38.8%	38.2%
High blood pressure	23.2%	20.6%
Asthma	13.0%	13.9%
Diabetes	5.8%	5.6%

Use of Medical Oxygen

- Multiple studies document growing recognition of special problems posed by people who continue to smoke while under treatment with medical oxygen. For example, from March 3, 1999, to November 30, 2000, 12 oxygen-therapy fires in Philadelphia caused three deaths and injured seven others. In addition, in a November 2003 safety brochure, the Massachusetts Office of the State Fire Marshal reported that "Since 1997, 16 people have died and 20 other individuals have suffered severe burns or smoke inhalation in fires involving people who were smoking while using home oxygen systems." (See Appendix C for a more detailed review of these studies and their findings.)

- Most ongoing fire incident databases do not provide for reporting of medical oxygen as a factor. However, data contained in FIDO (see Table B-14 in Appendix B) revealed that 7 percent of fatal victims who were the smokers whose smoking material ignited the fire were under treatment with medical oxygen.

Engaging in Risky Behavior

- According to the CDC, smokers defined as those who have smoked at least 100 cigarettes in their lifetimes were more likely than nonsmokers to not always have used a seatbelt when driving or riding (25.5 percent versus 19.7 percent). In addition, these smokers were more likely than nonsmokers to have driven after drinking (4.4 percent versus 2.9 percent).

- According to the CDC, smokers defined as everyday or someday smokers were more likely than nonsmokers to have not always used a seatbelt when driving or riding (30.4 percent versus 25.4 percent). However, these smokers were **less** likely than nonsmokers to have driven after drinking (5.8 percent versus 6.8 percent).

Summary of Points on Correlated Complicating Characteristics

Many of these characteristics make it more difficult and less likely that smoker behavior can be modified effectively to improve safety. For example, a person impaired by drugs or alcohol probably is less likely to act effectively in accordance with learned safer behavior. A person with disabilities is less likely to be able to so act. A person who tends to engage in risky behaviors generally is less likely to be susceptible to behavior change to reduce one particular risk.

Therefore, this set of findings, in combination with the earlier findings regarding the large share of victims who are intimate with ignition, has the following implications for strategy selection:

- For many fatal victims of smoking-material fires, protection strategies that operate after ignition are unlikely to be successful. Prevention strategies need to be emphasized for these fires.

- The smokers whose actions and omissions led to fatal smoking-material fires are more likely to have significant barriers to behavior change than the individuals who are involved with other types of fatal fires. Therefore, strategies that modify cigarettes and/or the commonly first ignited objects (e.g., upholstered furniture, mattress, bedding, clothing) need to be included in an effective program to reduce risk of death due to smoking-material fire.

- One of the seven educational messages recommended from this project is intended to address the heightened risks associated with several of the following characteristics that adversely affect alertness:

To prevent a deadly cigarette fire, you have to be alert. You won't be if you are sleepy, have been drinking, or have taken medicine or other drugs.

- The EMAC that developed the recommended messages agreed that the emerging problem of smoking and medical oxygen needs to have a separate message focused on it, which can be used in situations where the use of medical oxygen is likely:

Smoking should not be allowed in a home where oxygen is used.

Summary of Findings on Smoker Behaviors and Related Strategies

Decision to Smoke

The decision to smoke or to smoke regularly is correlated with (a) other risk-increasing behaviors (e.g., drinking, use of other drugs); (b) stress; (c) early addiction; and (d) several clusters of conditions that could be expected to create stress.[4,5,6,7,8,9,10,11,12] Identified clusters include the following:

- personal experience with violence, as a victim, close relative of a victim, or victimizer;

- poor relationship with parents (e.g., lack of parental support, lack of parental presence before and/or after school, lack of activities with parents, single-parent household, lack of family connectedness);

- poor situation at school, either poor relationships generally, poor performance generally (e.g., repeating a grade, low grade point average), or personal limitations (e.g., emotional, learning, or mobility disabilities) that could make school difficult both for learning and for social relationships; and

- attitudinal strength characteristics, including lack of self-esteem, lack of religiosity, and belief in early death.

It is possible that the other risk-increasing behaviors are not causes of smoking but are correlated results of the same cause that leads to smoking, e.g., willingness to accept risk in activities that give pleasure or taking pleasure from risk itself.

Smokers also are more likely to underestimate what the risk is, and teenage smokers are more likely to discount risks whose consequences are many years away.[13] Many assume that they will not experience the uncertain negative outcomes. One successful antismoking program in Australia used the phrase "every cigarette is doing you damage" to encourage audiences to think of the risk as immediate.[14,15]

Teenage smokers are more likely than teenage nonsmokers to be angry. Successful antismoking campaigns in Florida, California, and Massachusetts channel the anger of teenage smokers toward tobacco companies, based on the latter's alleged manipulation of teenagers through targeted marketing and alleged dishonesty and callousness regarding the health effects of cigarettes.[13,16]

Smokers tend to value freedom to smoke over risks to themselves but can regard risks to nonsmokers (e.g., secondhand smoke) as an effective argument against smoking.[17]

The decision to smoke is part of the sequence of smoker behaviors that can lead to a fire. This brief synopsis of results of antismoking programs focuses on findings related to smoker characteristics that could be useful in designing strategies for changing smoker behaviors to prevent fires. However, it was considered outside the scope of the project to develop strategies focused specifically on the decision to smoke. It was further determined, based on focus group research by Hager Sharp, the public relations firm, in another project funded by USFA, that strategies to change smoker behaviors would be more accepted by smokers if they did not appear to be intended to curtail smoking.[18]

Consequently, none of the recommended behaviors and associated educational messages are meant to address the decision to smoke, but by the same token, none are phrased so as to appear to endorse or explicitly accept the decision to smoke.

What Cigarettes Ignite

The majority of smoking-material home structure fires (55 percent in 1999 to 2001) and more than two-thirds of associated deaths (71 percent in 1999 to 2001) involve trash, mattress or bedding, or upholstered furniture as the item first ignited (Figure 7). Trash is a major contributor only to fire incidents and not to fatal fires.

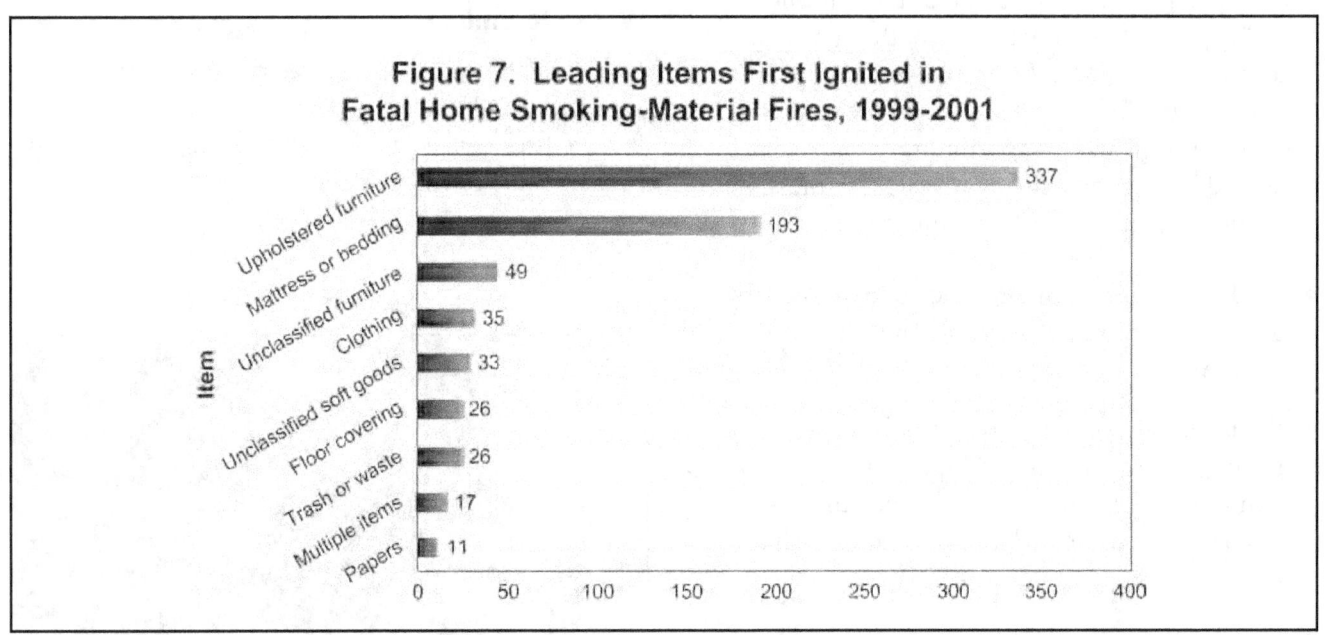

Figure 7. Leading Items First Ignited in Fatal Home Smoking-Material Fires, 1999-2001

Note: These are national estimates of fires reported to U.S. municipal fire departments. Fires reported only to Federal or State agencies or industrial fire brigades are excluded. National estimates are projections. Casualty and loss projections can be influenced heavily by the inclusion or exclusion of one unusually serious fire. Deaths are estimated to the nearest one. Figures reflect a proportional share of home structure smoking-material fires with item first ignited unknown. Figures do not include adjustment for fires coded as unclassified or unknown-type open flame or smoking-material or for confined fires.

Source: NFIRS and NFPA survey.

Both mattresses and upholstered furniture have been the subject of decades-long requirements, industry-based or government-based, respectively, to reduce ignitability by cigarettes. The long-term impact of these programs can be seen in the rising percentage of fatal home structure smoking-material fires that begin with ignition of something other than upholstered furniture, mattresses, or bedding. That percentage was 15 percent in 1980 to 1982, 20 percent in 1990 to 1992, and 29 percent in 2000 to 2002.

Product Choices and the Control of Burning Cigarettes

There are limits to the cigarette resistance one can build into potential items first ignited. When those limits are reached, two alternatives are to change the cigarette itself (discussed later) and to improve practices that keep the lit cigarette and potential items first ignited apart. Ashtrays are intended to be used to achieve that separation.

The characteristics of an effective ashtray have been described by many different terms in existing educational materials, but some of those terms (e.g., "large") were judged to be both vague and potentially inadequate. An ashtray is intended to provide a safe repository for ashes while a cigarette is being smoked and a safe temporary repository for ashes and butts after a cigarette has been smoked. This will happen if the ashtray minimizes

- the likelihood of a lit cigarette falling out of the ashtray (depth was deemed the most important feature);

- the likelihood of the ashtray itself overturning and spilling ashes, embers and butts onto potential combustibles ("sturdy" was deemed the best established term for what is needed in such an ashtray); and

- the likelihood of a hostile fire if ashes, embers, or butts fall outside the protected confines of the ashtray (a sturdy, hard-to-ignite surface for the ashtray was deemed the best way to describe what was needed).

Neither literature nor data could be found on the relative frequency of smokers' practices regarding ash disposal or butt disposal, whether fire ensued or not, or on the time required for butts and ashes, left alone but not actively dowsed, to become safe for disposal into ordinary trash containers. Therefore, it was not possible to determine the importance of dowsing as part of a responsible smoker's routine.

Two educational messages developed for this project address behaviors related to effective use of ashtrays and effective techniques for disposal of cigarettes after smoking:

Wherever you smoke, use deep, sturdy ashtrays. Ashtrays should be set on something sturdy and hard to ignite, like an end table.

Before you throw out butts and ashes, make sure they are out, and dowsing in water or sand is the best way to do that.

The second message emphasizes the need for safe disposal and cites dowsing as a preferred method but leaves flexibility as to the choice of means.

Where to Smoke

Nearly half of all smoking-material home structure fires and roughly three-fourths of associated deaths involve fires that begin in the bedroom or the living room, family room, or den.

Home Structure Fires Ignited by Smoking Materials
Annual Averages 1999 to 2001, by Area of Origin

Area of Fire Origin	Fires		Civilian Deaths		Civilian Injuries		Direct Property Damage (in Millions)	
Bedroom	6,800	29%	270	34%	860	46%	$109	31%
Living room, family room, or den	3,600	15%	330	43%	490	26%	$79	23%
Exterior balcony or unenclosed porch	1,800	8%	0	0%	30	2%	$28	8%
Trash chute or container	1,700	7%	0	0%	20	1%	$14	4%
Kitchen	1,600	7%	30	3%	70	4%	$19	5%
Bathroom	1,100	4%	10	1%	50	3%	$6	2%
Exterior wall surface	900	4%	0	0%	20	1%	$7	2%
Garage or carport	600	3%	0	0%	30	1%	$11	3%
Exterior stairway, ramp, or fire escape	400	2%	0	0%	0	0%	$1	0%
Substructure area or crawl space	400	2%	10	1%	20	1%	$5	1%
Laundry area	400	2%	0	0%	20	1%	$4	1%
Courtyard, patio, or terrace	400	2%	0	0%	10	0%	$4	1%
Hallway or corridor	300	1%	0	0%	10	0%	$1	0%
Dining room	300	1%	30	4%	40	2%	$7	2%
Other known area of origin	3,500	15%	100	13%	210	11%	$53	15%
Total	23,700	100%	780	100%	1,870	100%	$349	100%

Note: These are national estimates of fires reported to U.S. municipal fire departments. Fires reported only to Federal or State agencies or industrial fire brigades are excluded. National estimates are projections. Casualty and loss projections can be influenced heavily by the inclusion or exclusion of one unusually serious fire. Fires are shown to the nearest hundred, deaths and injuries to the nearest ten, and property damage, unadjusted for inflation, to the nearest million dollars. Figures reflect a proportional share of home structure smoking-material fires where the item first ignited is unknown. Figures do not include adjustment for fires coded as unclassified or unknown-type open flame or smoking material or for confined fires. Totals may not equal sums because of rounding.

Source: NFIRS and NFPA survey.

Available data do not permit calculation of the risk of fatal cigarette fires relative to time spent smoking, distinguishing different rooms of a home. Therefore, although it is rare for fatal home smoking fires to begin in such rooms as the kitchen (3 percent) or bathroom (1 percent), it is not possible to say that an hour spent smoking in those rooms carries less risk than an hour spent smoking in the bedroom or living room.

There is some indirect evidence of the relative safety of smoking outdoors. In 1994 to 1998, in the months of December through February, the rate of deaths per 100 reported smoking-material home structure fires was much higher--4.9 in December

through February and 2.8 in the other months, or 75 percent higher in winter. Winter is when going outdoors to smoke is most difficult and, therefore, least likely to happen.

Based on this evidence, one of the principal educational messages focuses on where people smoke:

If you smoke, smoke outside.

Because some may be concerned over smoker resistance to this behavior change, it is useful to note evidence that, in many cases, rules like these may actually be seen as **reinforcing** established household rules rather than reinventing behavior.

- According to the CDC, approximately half (53 percent) of smokers defined as those who have smoked at least 100 cigarettes in their lifetimes have rules against indoor smoking at home, while four out of five (81 percent) nonsmokers have such rules. Furthermore, 15 percent of smokers and 5 percent of nonsmokers have rules limiting where or when smoking is permitted inside the home.

- According to the CDC, approximately one-fourth (27 percent) of smokers defined as everyday or someday smokers have rules against indoor smoking at home, while approximately half (52 percent) of nonsmokers have such rules. Furthermore, 24 percent of smokers and 22 percent of nonsmokers have rules limiting where or when smoking is permitted inside the home.

- Rules on where to smoke were more common when children and a nonsmoking partner were present.[17]

The Philadelphia (PA) Fire Department has an innovative outreach program centered on this message. One of their firefighters, Rodney Jean-Jacques, doubles as a hip hop artist under the name of Cal Akbar. He has developed a song and video titled *Take It Outside*. More information is available from Lt. Michael Grant of the Fire Prevention Division (e-mail michael.grant@phila.gov).

What to Smoke

As discussed, there are limits to the cigarette resistance one can build into potential items first ignited. When those limits are reached, another alternative is to change the cigarette itself. Two major Federal-government studies of the potential risk reduction from a reduced ignition-strength cigarette reached the following conclusions:

- A reduced ignition-strength cigarette is technically feasible.

- A standard test of cigarette ignition strength is technically feasible. (A standard test was developed by ASTM Committee E05, Fire Standards, and approved in 2002 as ASTM E2187. The name of the organization is now just the initials, ASTM, which originally stood for American Society for Testing and Materials.)

See Appendix D for a more detailed and referenced discussion of this issue.

For several years, smokers have been able to purchase cigarettes with banded paper, a technology shown to produce reduced ignition strength without higher costs or adverse side effects (such as higher toxicity). However, availability has been limited, as has publicity, and brand loyalty remains a powerful factor in smoker choice. It is not yet possible to identify factors in the decisions of smokers to select or avoid reduced-ignition-strength cigarettes. Therefore, the project developed a recommended educational message that simply draws attention to the existence and effectiveness of the technology:

If you smoke, choose fire-safe* cigarettes. They are less likely to cause fires.

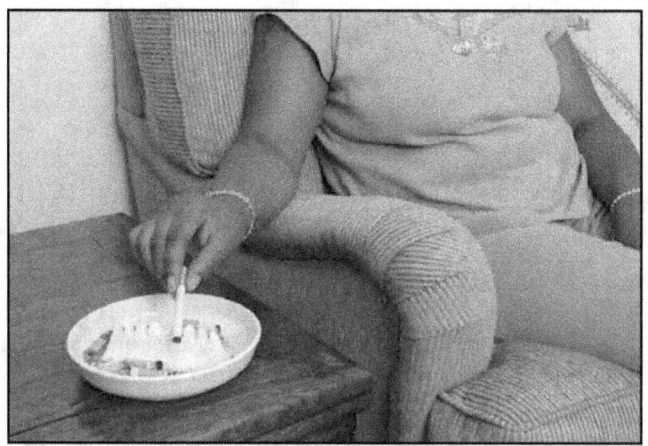

Smoker Characteristics Related to Product Choice Behaviors

Several parts of the analysis pointed to product choice behaviors that could significantly reduce risk. These include the choice of reduced ignition-strength cigarettes, the choice of more cigarette-resistant upholstered furniture and mattresses, and even the choice of better ashtrays and more suitable furnishings on which to place those ashtrays.

---

*The cigarettes in question are not really "fire-safe" in the implied absolute sense, and places like New York State, which have adopted or considered requirements for reduced ignition strength are careful to avoid the term "fire-safe." Therefore, it may be desirable to revisit the wording of this message.

While most involve features not found as much or at all in older products, none of these product choices appear to involve higher costs. However, according to the CDC risk factor database, smoking is correlated with poverty and with lower educational achievement. These two factors could hinder any attempt to redirect product choices and make it more likely that old products, made before features of greater safety were required or invented, will be in place. For example, poor households may be more likely to purchase used products and avoid replacing old products that are still serviceable.

According to the CDC, smokers defined as those who have smoked at least 100 cigarettes in their lifetimes were more likely than nonsmokers to have less than a high school education (13.2 percent versus 9.4 percent) and to have household income of less than $20,000 per year (22.6 percent versus 19.3 percent).

According to the CDC, smokers defined as everyday or someday smokers were more likely than nonsmokers to have less than a high school education (16.5 percent versus 11.5 percent) and to have household income of less than $20,000 per year (27.4 percent versus 24.2 percent)

Presence of Others as a Factor in Strategy Design

Some strategies to change smoker behaviors or the consequences of those behaviors depend on the influence of other persons with regard to whether, what, where, when, and how to smoke. In addition, they can provide and help maintain an environment that is more forgiving of stray cigarette butts and ashes.

According to FIDO, the smoker whose smoking materials ignited the fire was the only person present in just over half (54 percent) of fatal cigarette fires. In these fires, however, smokers may not have lived alone and may have been influenced by others.

According to FIDO, one fatal victim in four (24 percent) is **not** the smoker whose cigarette started the fire. Therefore, if others are present, they have both a direct and an indirect stake in taking action to prevent hostile fires from taking place. These victims are present in some types of rooms more than others:

- When fire began in the **bedroom**, only one of every nine victims (11 percent) with known location were not the smokers whose smoking began the fire.

- When fire began in the **living room, family room, or den**, nearly one-third of victims (31 percent) with known location were **not** the smokers whose smoking began the fire.

- When fire began in **any other room** or area, more than one-fourth of victims (28 percent) with known location were **not** the smokers whose smoking began the fire.

The relationships of these victims to the smokers are useful to know, as the relationships may bear on the victims' willingness and ability to serve as "watchers" for the smokers. The relationships also may bear on the willingness of the smokers to accept help or advice from these others. According to FIDO, the following statistics describe the relationships to the smokers of the fatal victims who were not those smokers:

- Thirty-four percent had the smokers as parents (not all the victims were under age 18);

- Twenty-five percent were neighbors (often from other apartment units in the same building) or friends;

- Fourteen percent were spouses or partners;

- Thirteen percent were parents of the smokers; and

- Fourteen percent had other relationships (e.g., sibling, niece or nephew, uncle or aunt, roommate, passerby).

The spouses, partners, and parents (one-fourth of the total when combined) are reasonably likely to be present in the home most of the time. This means they are likely to be present in the home most of the time and, as a result, would be available to act in support of the safety of the household from smoking-material fires. Also, as spouse, parent, or partner, they are likely to have some influence on the behavior of their partners or children who are smokers.

The children (one-third of the total) are also reasonably likely to be present but may be less able and willing to influence their smoker parents. The other victims (two-fifths of the total when combined) are unlikely to be present to provide such influence and lack the kind of close relationship that would make such influence seem appropriate.

According to FIDO, 2 percent of total fatal victims were transients who were not legal occupants of the structures in which the fire began. These might be the smokers or others. This would be a particularly difficult group to educate in safer behaviors.

According to FIDO, fatal victims who were **not** the smoker whose cigarette started the fire (one-fourth of total victims) are much less likely to have been in the room where the fire began, let alone close enough to the point of fire origin to be called "intimate with ignition."

- Of the children killed by smoking-related fires, 14 percent were killed in a living room, family room, or den, always by a fire that began in the same room. The other 86 percent were killed in a bedroom, always by a fire that began in a different room. That different room was usually the living room, family room, or den.

- Of the neighbors and friends killed by smoking-related fires, 20 percent were killed in a living room, family room, or den by a fire that began in the same room. None of the other 80 percent were killed in the room where fire began.

- Of the spouses and partners killed by smoking-related fires, only 22 percent were killed in the room where fire began. That room was the bedroom for half of those victims (11 percent of all spouse and partner victims) and the living room, family room, or den for the other half.

- Of the parents of smokers killed by smoking-related fires, none were killed in the room where fire began.

- Only 13 percent of all nonsmoker fatal victims combined were injured in the room where the fire began. For that 13 percent of victims, the room where the fire began and they were fatally injured was nearly always a living room, family room, or den.

- Therefore, 87 percent of all nonsmoker fatal victims were injured in a different room from the room where fire began.

The statistics cited in this section have some implications for behavioral strategies:

- The majority of smoker victims had no one else in the housing unit when they were fatally injured.

- For those smokers who had someone else present, our only detailed information is on those others who were also fatal victims.

- Based on their relationships to the smokers, many of those others were unlikely to be present much of the time (e.g., neighbors, friends) and some might have difficulty in influencing the smoker.

- Most of the nonsmoker victims were not in a position to see the fire start. To act as "watchers," they would have to take some kind of proactive action, not just maintain a heightened alertness to fire outbreaks.

One of the recommended educational messages focuses on proactive actions like checking for butts and ashes after smoking has taken place. This is a behavior that nonsmokers can do as easily as smokers, and if nonsmokers are less likely than smokers to have been drinking, as the CDC database suggests, then nonsmokers may be a more reliable and available group to focus this message on:

Check under furniture cushions and in other places people smoke for cigarette butts that may have fallen out of sight.

Victims who are not fatally injured in the room where fire begins also are more likely to be savable by fire protection--working smoke alarms, fire sprinklers, compartmentation--because they tend to be farther from the fire and so have time for fire protection features and systems to work. This means the nonsmoker victims are more savable by these means than the smoker victims.

Converting Research Findings into Behavioral Strategies

The research findings characterized the size and characteristics of the smoking-material fire problem. Most of these findings were useful in helping define what behaviors needed to change in order to reduce the smoking-material fire problem. However, the findings did not indicate what specific behaviors would be the best choices to target as the new behaviors by smokers and those around them.

NFPA's new EMAC was created to provide a diverse and knowledgeable volunteer base for consensus recommendations on educational messaging. A day was set aside for them to develop and agree on messages, with the research findings as input.

The project recommends the use of four general messages, two more specific messages for particular audiences, and a seventh message to be used when space or time permit. The four general messages are as follows:

If you smoke, smoke outside.

Wherever you smoke, use deep, sturdy ashtrays. Ashtrays should be set on something sturdy and hard to ignite, like an end table.

Before you throw out butts and ashes, make sure they are out, and dowsing in water or sand is the best way to do that.

Check under furniture cushions and in other places people smoke for cigarette butts that may have fallen out of sight.

The two specific messages:

Smoking should not be allowed in a home where oxygen is used.

If you smoke, choose fire-safe cigarettes. They are less likely to cause fires.

And the seventh message:

To prevent a deadly cigarette fire, you have to be alert. You won't be if you are sleepy, have been drinking, or have taken medicine or other drugs.

One of the principles used by the EMAC was that behavioral-change messages are more likely to be accepted and acted upon if they are stated positively ("do this") rather than negatively ("don't do this"). For the last message shown above, this principle required a bit of a stretch in which the main message, stated positively, was also rather general. The second sentence contained detail necessary for clarity. That detail was unavoidably negative in form, but the sentence was phrased as an embellishment of a positive message rather than as a negative, imperative statement.

Making Behavior Change More Likely

The research findings did not identify any strong results that would point to best strategies to try to change behaviors in the directions selected by the EMAC. Instead, the project team reviewed the six questions in a popular model of health belief formation for their implications.[18] The project team also drew insight from focus group work with smokers in a parallel project conducted for the USFA by Hager Sharp.[19]

What Factual Evidence Needs to Accompany the Educational Message?

The first question in the Health Belief model asks about the target individual's existing assessment of likelihood of harm targeted by the strategies, which in this context would be the experience of having a smoking-related fire and especially suffering an injury, particularly a fatal injury, in such a fire. The second question asks about the target individual's assessment of the seriousness of such harm.

Hager Sharp found that smokers and nonsmokers were unaware of the fire losses. However, Hager Sharp found no resistance to the idea that nearly 1,000 fire deaths a year was very serious.

Hager Sharp also found that smokers tend to believe they are aware of the rules of fire-safe smoking behavior, and there was a clear suggestion that many smokers therefore believe that their own probability of fire loss is less than the national loss figures would suggest.

Teenagers considering whether to smoke tend not to be influenced by long-term health considerations.

The NFPA EMAC, which developed the recommended educational messages around which our strategies are organized, also recommended that **facts about the size and severity of the smoking-material fire problem be communicated with the educational messages.**

When assembling such a fact summary, there are choices to be made on what types and examples of harm to emphasize. A case for high severity might emphasize any or all of the following:

- very serious nonfatal injuries (where the long-term costs of care and the disfiguring nature of the injuries may give them even more impact than deaths);

- very broad definitions of relevant injuries (e.g., unreported injuries);

- very broad definitions of smoking-related fires (e.g., fires involving lighting implements but not smoking); and

- very inclusive calculations of the costs of injuries (e.g., including pain and suffering as monetized in tort litigation).

However, it is possible to create a backlash by making the problem of concern appear too terrible. Studies of effective strategies in safety education not limited to fire safety have concluded that images of serious harm can induce a coping reaction of avoiding the images rather than changing the targeted behavior.

Our recommended approach emphasizes the use of the core facts in plain form.

What Detailed Guidance Should Accompany the Educational Message?

It is recommended that the educational messages be written to provide additional information on **how** to be careful with cigarettes.

For example, smokers may assume they know the importance of using ashtrays to control burning cigarettes safely, but they may be open to information on the kinds of ashtrays (i.e., deep, sturdy) to use and the ways to use them (e.g., on a sturdy foundation that will not burn).

Additionally, smokers may assume they know the importance of not dropping cigarette butts or ashes onto or into soft furnishings, but they may be open to the idea of a confirmatory check on those furnishings to make sure.

This argues for the use of pictures (e.g., good versus bad ashtrays, where and how to look for butts or ashes) in materials disseminating the messages.

What Side Effects Could Be of Concern for the Recommended Strategies?

One of the Health Benefit model questions has to do with the target audience's perception of adverse side effects from a recommended strategy.

A number of adverse side effects have been alleged for the reduced-ignition-strength cigarette. These claims and refutations of these claims have been playing out in testimony about proposed bills to require exclusive use of such cigarettes. Appendix

D has a bit more information on the types of side effects being cited.

One of our behavioral strategy recommendations is to move most home smoking outdoors. Changing where or how one smokes while at home means giving up the comfort of a familiar experience and thinking consciously about risks and safety more of the time. These could be perceived as adverse side effects. Smoking outdoors may mean a less comfortable environment (not climate controlled) and less opportunity to socialize with others who are not smoking, to use the entertainment equipment of the home, and so forth. For all these reasons, we predict that the effort to move outdoors will encounter resistance, and we include that recommendation primarily because the CDC risk factor database indicates that a large percentage of smokers already follow such rules in their homes.

Keeping Messages Short, Simple, and Clear

Hager Sharp found resistance to such terms of art as "abandon," "extinguish," and "discard." All of these are terms that could slip through the usual education filter of avoiding words with more than three syllables.

This consideration was a reason for keeping the number of messages down to seven and identifying just four messages as primary messages for all audiences.

Would Humorous Messages Work in This Context?

Hager Sharp explored the use of humor in messages as a means of reducing resistance from smokers. Hager Sharp found humor was well received, but one person's humor can be another person's disrespect. Humor always has a target, and it is not clear what would constitute a safe but effective target for humor in this context.

For example, consider the messages with humor that Hager Sharp tested:

"Don't get in bed with a butt."

"Don't barbecue in bed. Don't smoke either; it's just as deadly."

The first message relies on double entendre. There are people who have problems with humor based on veiled sexual references, even when it is as mild as this sample.

The second message relies on analogy. Very few analogies are universally convincing or even understood, and some can be distracting.

The history of advertising is replete with humor-tinged ads that were widely enjoyed but not particularly effective in changing behavior, specifically, increasing purchases. Humor that fails can well be worse than no humor at all.

This project does not recommend the use of humor in formulating messages for this fire problem.

National Strategies for Implementation of Report Findings

See Appendix E for the application of the recommended messages to existing USFA educational materials.

The implementation of the project results should be pursued through integration of the messages into established public fire safety programs and campaigns.

Training and Equipping the Educators

Two PowerPoint® presentations have been developed from the project results. The first such presentation is targeted at smokers and others whose behavior we seek to change. (See Attachment I.) In order to gain clarity, many of the messages are accompanied by appropriate photographs. The second such presentation is targeted at educators and provides more of the details of the project and the thinking behind the messages. (See Attachment II.)

We recommend that these two presentations be made available free on the USFA Web site. They also should be used proactively in fire safety educator training sessions and conferences.

At some point, it would be useful to provide public service announcements in various media for national or local deployment. We recommend that model public service announcements be developed from the results of this project with the help of public relations experts, such as Hager Sharp who have already done some work for the U.S. Fire Administration on the subject of smoking-material fires, as cited earlier.

Standardization of Educational Messages

The recommended messages developed for this project emerged from an extension of NFPA's new EMAC. This group is well equipped and well positioned to provide ongoing maintenance of these messages, and we recommend that they be asked to do so.

We also recommend that these messages be subjected to focus group and pilot project evaluation. Findings from those evaluations will provide a more detailed substantive basis for the selection and specific wording of the messages.

Implementation of Findings Beyond USFA

This report and these messages should be proactively distributed to other developers and distributors of fire safety public education materials, whether focused on fire safety or included in larger packages on unintentional injuries or health generally. We recommend that this distribution include strong encouragement to others to incorporate these findings into their materials.

NFPA has already taken the initiative to begin implementing these findings in its public education materials, as they come up for routine revision.

Future Research

There were a number of points in the message development process where better technical information would have supported better or more confident choices. The following are topics where additional research could lead directly to better messages:

- **Is there a relatively safe place to smoke indoors?** Some members of the messaging committee wanted to recommend that smokers use the kitchen if they cannot or will not smoke outdoors. Ultimately, that message was not included because too many questions about the risk associated with smoking in the kitchen could not be answered.

Smoking-material fires and especially fatal fires are much more common in bedrooms, living rooms, family rooms, and dens than in kitchens, but it is not known how the difference in fire frequencies compares with the difference in time spent smoking. The question remains whether fires are less frequent in kitchens primarily because people do not smoke in the kitchen very often, as compared to other rooms.

In addition, the argument was made that vulnerable soft furnishings are less common in kitchens. No information was available, however, to confirm the magnitude of differences in vulnerability of contents to cigarette ignition and to subsequent fire growth.

This research project would involve assembling more complete data on the fuel loading in kitchens versus other rooms and on time spent smoking in kitchens versus other rooms. The same question has been raised regarding bathrooms.

- **How big a problem is smoking around medical oxygen?** The available data were taken from a small number of jurisdictions and were neither large enough nor representative enough to support national estimates. A research project to better quantify the magnitude of this problem would be useful as an indication of the priority that should be attached to this part of the smoking-material fire problem.

- **What is required to clear an area of medical oxygen before smoking?** The recommended message is very conservative and states there should be no smoking by anyone in a home where medical oxygen is used. The project team was unable to find any substantiated guidance on what is required, beyond turning off the flow of oxygen, to make a room or area safe for smoking. Because oxygen can be absorbed by clothing or other fabric in the area, the answer is not obvious. A research project could provide a basis for less conservative but still effective guidance on safe practices.

- What behaviors would people be willing to perform to achieve safety when disposing of butts and ashes? The project team went back and forth regarding the strength of the recommendation to use immersion in water or sand to assure that ashes and butts were safe for disposal. A research project could examine the ignition potential in butts and ashes based on other methods, such as stubbing out the cigarette and leaving it in the ashtray for a defined period of time. A research project also could develop better information on likely smoker resistance to messages that would require behaviors as time-consuming as dowsing in water or sand. The same project could look into variations of the message regarding inspection of couch cushions and other hidden sites for stray butts or ashes. More detailed, substantiated guidance on how often, when, and how this can best be done would be helpful.

- What approach works best in health care facilities? A decade or more ago, health care facilities began taking a more aggressive approach to smoking in their facilities. Three different approaches emerged.

One approach was a complete ban on smoking by anyone. This is clearly the most effective approach if the ban is observed, but concerns were raised over the risks posed by "hidden smoking," in which the lit cigarette may be held closer to soft furnishings while being hidden from health care professionals.

The second approach was restriction of patient smoking to designated areas. It is not known how well those areas have been secured from risk of cigarette-initiated fire or how successful facilities have been in achieving compliance.

The third approach was a requirement that patients be attended by medical personnel when they smoke. Again, it is not known how successful facilities using this approach have been in achieving compliance.

A research project could determine the popularity and history of effectiveness of each of these approaches.

References

1. Holborn, Paul G. *The Real Fire Library: Analysis of Fatal Fires 1996-2000*. London Fire Brigade--Fire Investigation Group, London Fire and Emergency Planning Authority, Nov. 2001.

2. Quillan, Thomas C. *An Analysis of Civilian Fire Deaths in Tallahassee (Leon County) Florida: 1983-1994, Strategic Analysis of Community Risk Reduction*. Applied research project submitted to the National Fire Academy as part of the Executive Fire Officer Program, Jan. 1995.

3. U.S. Fire Administration/National Fire Data Center. *Case Study: Contribution of Alcohol to Fire Fatalities in Minnesota*. Topical Fire Research Series, Vol. 3, Issue 4, July 2003, online at http://www.usfa.fema.gov/inside-usfa/nfdc/pubs/tfrs.shtm

4. Blum, Robert W., Anne Kelly, and Marjorie Ireland. "Health-Risk Behaviors and Protective Factors among Adolescents with Mobility Impairments and Learning and Emotional Disabilities." *Journal of Adolescent Health*, 28 (2001): 481-490.

5. Barbeau, Elizabeth M., Nancy Krieger, and Mah-Jabeen Soobader. "Working Class Matters: Socioeconomic Disadvantage, Race/Ethnicity, Gender, and Smoking in NHIS 2000." *American Journal of Public Health* 94, no. 2 (Feb. 2004): 269-278.

6. Minnesota Department of Health, Blue Cross and Blue Shield of Minnesota, Minnesota Partnership for Action Against Tobacco, and University of Minnesota.

Patterns of Smoking among Minnesota's Young Adults. Jan. 2004, online at http://www.health.state.mn.us/divs/hpcd/tpc/TobaccoReports.html

7. Office on Smoking and Health, Division of Adolescent and School Health, National Center for Chronic Disease Prevention and Health Promotion. "Cigarette Use among High School Students--United States, 1991-2003." *MMWR Weekly* 53, no. 23 (2004): 499-502, online at http://www.cdc.gov/mmwr/preview/mmwrhtml/mm5323a1.htm

8. Rode, Pete. *Youth Risk Behavior and Social Factors Associated with Smoking Cigarettes--Results from the 2001 Minnesota Student Survey; 6th, 9th and 12th Grade Students*. Minnesota Department of Health, Center for Health Statistics, Aug. 9, 2002, online at http://www.mnschoolhealth.com/data.html

9. Rode, Pete. *Teens and Tobacco in Minnesota--Results from the Minnesota Youth Tobacco Survey*. Dec. 2001, online at http://www.health.state.mn.us/divs/chs/data/youthtob.pdf

10. Harrell, Joanne, S., Shrikant I. Bangdiwala, Shibing Deng, Julie P. Webb, and Chyrise Bradley. "Smoking Initiation in Youth: The Roles of Gender, Race, Socioeconomics, and Developmental Status." *Journal of Adolescent Health* 23 (1998): 271-279.

11. Siqueira, Lorena, Marguerite Diab, Carol Bodian, and Linda Rolnitzky. "Adolescents Becoming Smokers: The Roles of Stress and Coping Methods." *Journal of Adolescent Health* 27 (2000): 399-408.

12. Viahov, David, Sandro Galea, Jennifer Ahern, Heidi Resnick, and Dean Kilpatrick. "Sustained Increased Consumption of Cigarettes, Alcohol, and Marijuana among Manhattan Residents after September 11, 2001." *American Journal of Public Health* 94, no. 2 (Feb. 2004): 253-254.

13. Goldman, Lisa K., and Glantz, Stanton A. "Evaluation of Antismoking Advertising Campaigns." *JAMA* 279, no. 10 (1998): 772-777.

14. Hill, David, Simon Chapman, and Robert Donovan. "The Return of Scare Tactics." *Tobacco Control* 7 (1998):5-8, online at http://tc.bmjjournals.com/cgi/reprint/7/1/5.pdf

15. Wakefield, M., J. Freeman, and R. Donovan. "Recall and Response of Smokers and Recent Quitters to the Australian National Tobacco Campaign." *Tobacco Control* 12 (2003): ii15-ii22, online at http://tc.bmjjournals.com/cgi/content/abstract/12/suppl_2/ii15

16. Niederdeppe, Jeff, Matthew Farrelly, and M. Lyndon Haviland. "Confirming 'Truth': More Evidence of a Successful Tobacco Countermarketing Campaign in Florida." *American Journal of Public Health*, 94, no. 2 (2004): 255-257.

17. Okah, Felix A., Won S. Choi, Kolawole S. Okuyemi, and Jaspit. S. Ahluwalia. "Effects of Children on Home Smoking Restrictions by Inner City Smokers." *Pediatrics* 109, no. 2 (2002): 244-249.

18. K. Glanz, B.K. Rimer, and F.M. Lewis, editors, *Health Behavior and Health Education: Theory, Research and Practice*, 3rd ed., San Francisco: Jossey-Bass, 2002.

19. Hager Sharp. Unpublished final report to U.S. Fire Administration on smoker receptiveness to educational messages, September 30, 2004.

Appendix A

Characteristics of Databases Used in Study

National Fire Incident Reporting System (NFIRS)

The USFA's NFIRS database is the most representative national fire database providing detailed information on individual fires and casualties. Nearly all national estimates of specific aspects of the U.S. fire problem begin with NFIRS. About a third of U.S. fire departments--working through their respective States-- participate in NFIRS, which receives reports on an estimated one-third to one-half of each year's fires.

NFPA and most other users of NFIRS combine it with the NFPA survey to produce the best "national estimates" of the specific characteristics of the Nation's fire problem.

NFIRS was established in the mid-1970's with only a couple of States participating. By 1980, there was sufficient representation from all regions to have high confidence in its representativeness. Thus, most analyses of NFIRS go back only to 1980. NFIRS usually is released on computer tape in the spring of the second year following the year it covers.

Fire Incident Data Organization (FIDO)

Many questions of technical interest require a level of detail beyond that available through NFIRS or the NFPA survey. For these, the best approach often is to use exploratory data from a database with sufficient validated detail to be useful, even though that database may not be representative of overall U.S. fire experience. The largest such database, excluding proprietary insurance-industry databases that focus on commercial properties and financial losses, is NFPA's Fire Incident Data Organization (FIDO).

FIDO is an incident-based database that combines information from fire departments, insurance companies, Federal and State safety agencies, and news sources. NFPA designed and operates FIDO to provide the best examples of fires demonstrating very specific phenomena of high technical interest. Examples include fires with information on the performance of specific types of fire protection systems and features.

When FIDO is not representative of all reported fires, it is primarily because it favors larger, more severe fires, such as fires involving five or more deaths or at least $5 million in property damage.

FIDO has operated since 1971 as a computerized index to NFPA's well-documented fires. FIDO contains data on almost 92,000 fires and now adds roughly 1,500 fires a year. There are up to 113 separate datum elements on each incident, and coverage extends to fires of interest around the world.

In 1997 to 1998, the FIDO data collection instructions were set to include all fatal fires, and the FIDO database is considered representative of all U.S. fatal fires for those years.

Many of the datum elements sought were never or almost never recorded, including detailed information on exactly where the fire-causing cigarette was located when it ignited a fire and where the smoker habitually smoked while at home. For most incidents, it was possible to identify persons present when fire occurred based on these characteristics: (a) whether they were the smoker whose cigarette started the fire or not; (b) relationship to the smoker if not the smoker; (c) whether the person was fatally injured, nonfatally injured, or uninjured; (d) personal limitations that could have influenced the ignition or reaction to fire after ignition, including distinguishing alcohol from drug impairment, identifying use of medical oxygen, and identifying up to three different limitations per person; (e) whether the person was in the room of fire origin and what type of room he or she was in.

27

For this project, data were extracted from 300 qualified 1997 to 1998 incidents, with records on 477 individuals, including 389 deaths.

CDC Behavioral Risk Factor Surveillance System

The CDC conducts an annual random-sample telephone survey called the Behavioral Risk Factor Surveillance System. The latest year of available data for this study was 2002, for which 247,964 interviews were conducted and processed.

Every cell used in analysis had at least 400 cases except for the following: yes to smoking a pipe now (45 to 1,005 cases, depending on cigarette-smoker status); most responses regarding bidis (minimum of 20 cases for current bidi smokers who are not current cigarette smokers); some answers within the not-always responses for nonsmokers regarding seatbelt use (analysis only contrasted always with not-always, and both of these had at least 400 cases); and some answers within the "allowed" group regarding where smoking is allowed (analysis only contrasted allowed versus not-allowed, and both of these had at least 400 cases).

Appendix B

Additional Statistics on Characteristics of the Victims of Smoking-Material Fires Related to Vulnerability

Table B-1. Ages of Victims of Smoking-Material Fires in U.S. Home Structures 1994-1998 Annual Average of Fire Deaths and Injuries

Age	1996 Population (in Millions)		Civilian Deaths		Death Rate per Million Persons	Civilian Injuries		Injury Rate per Million People
5 and under	23.3	(8.8%)	49	(5.2%)	2.1	72	(3.3%)	3.1
6 - 9	15.4	(5.8%)	11	(1.1%)	0.7	35	(1.6%)	2.3
10 - 19	37.6	(14.2%)	20	(2.1%)	0.5	173	(7.8%)	4.6
20 - 29	36.6	(13.8%)	64	(6.9%)	1.8	332	(15.0%)	9.1
30 - 49	83.2	(31.4%)	257	(27.4%)	3.1	808	(36.5%)	9.7
50 - 64	35.3	(13.3%)	199	(21.2%)	5.6	323	(14.5%)	9.1
65 - 74	18.7	(7.0%)	144	(15.3%)	7.7	227	(10.2%)	12.2
75 and over	15.2	(5.7%)	194	(20.7%)	12.7	246	(11.1%)	16.2
Total	265.3	(100.0%)	937	(100.0%)	3.5	2,217	(100.0%)	8.4

Note: These are national estimates of fires reported to U.S. municipal fire departments. Fires reported only to Federal or State agencies or industrial fire brigades are excluded. National estimates are projections. Casualty and loss projections can be influenced heavily by the inclusion or exclusion of one unusually serious fire. Statistics include a proportional allocation of fires with unknown form of heat of ignition and of smoking-related fires where the age of the victim was unknown. Civilian deaths and injuries are rounded to the nearest one. Totals may not equal sums because of rounding.

Sources: National estimates based on NFIRS and NFPA survey, and Statistical Abstract of the United States, Washington, DC: U.S. Department of Commerce, Bureau of the Census, 1997.

Table B-2. Age by Sex of Victims of Smoking-Material Fires in Home Structure Fires Reported to U.S. Fire Departments 1994-1998 Annual Average of Fire Deaths and Injuries

Age	1996 Population (in Millions)	Civilian Deaths	Death Rate per Million People	Civilian Injuries	Injury Rate per Million People
Male					
5 or younger	11.9	24	2.0	42	3.5
6-9	7.9	5	0.6	16	2.0
10-19	19.4	16	0.8	107	5.5
20-29	18.5	46	2.5	208	11.2
30-49	41.3	185	4.5	523	12.7
50-64	16.9	127	7.5	210	12.4
65-74	8.3	75	9.0	118	14.1
75 and over	5.6	87	15.7	86	15.5
Total	**129.8**	**566**	**4.4**	**1,310**	**10.1**
Female					
5 or younger	11.4	25	2.2	30	2.7
6-9	7.5	6	0.8	19	2.6
10-19	18.3	4	0.2	66	3.6
20-29	18.0	18	1.0	124	6.9
30-49	41.9	72	1.7	285	6.8
50-64	18.4	71	3.9	113	6.1
65-74	10.3	69	6.7	110	10.6
75 and over	9.6	106	11.0	160	16.6
Total	**135.5**	**371**	**2.7**	**907**	**6.7**

Note: These are national estimates of fires reported to U.S. municipal fire departments. Fires reported only to Federal or State agencies or industrial fire brigades are excluded. National estimates are projections. Casualty and loss projections can be influenced heavily by the inclusion or exclusion of one unusually serious fire. Statistics include a proportional allocation of fires with unknown form of heat of ignition and of smoking-related fires where the age or gender of the victim was unknown. Civilian deaths and injuries are rounded to the nearest one. Totals may not equal sums because of rounding.

Sources: National estimates based on NFIRS and NFPA survey; and Statistical Abstract of the United States, 1997, Washington: U.S. Department of Commerce, Bureau of the Census.

Table B-3. Victim Location at Ignition in Home Fire Casualties, Smoking-Related vs. Other Causes vs. All Causes 1994-1998 Structure Fires Reported to U.S. Fire Departments

A. Civilian Deaths

Location at Ignition	Smoking	Other Cause	All Causes
Intimate with ignition	29.1%	10.9%	15.8%
Not intimate, in room of origin	28.0%	23.1%	24.4%
Not in room, on floor of origin	28.2%	32.5%	31.4%
Not on floor, in building of origin	13.0%	31.9%	26.8%
Outside of building of origin	0.7%	0.8%	0.8%
Unclassified	1.0%	0.8%	0.8%
Total	**100.0%**	**100.0%**	**100.0%**

B. Civilian Injuries

Location at Ignition	Smoking	Other Cause	All Causes
Intimate with ignition	20.7%	14.9%	15.6%
Not intimate, in room of origin	25.6%	27.3%	27.1%
Not in room, on floor of origin	22.0%	25.3%	24.9%
Not on floor, in building of origin	24.2%	26.2%	25.9%
Outside of building of origin	5.7%	4.3%	4.5%
Unclassified	1.8%	2.0%	2.0%
Total	**100.0%**	**100.0%**	**100.0%**

Note: These are national estimates of fires reported to U.S. municipal fire departments. Fires reported only to Federal or State agencies or industrial fire brigades are excluded. National estimates are projections. Casualty and loss projections can be influenced heavily by the inclusion or exclusion of one unusually serious fire. Statistics include a proportional allocation of fires where the victim location was unknown.

Source: National estimates based on NFIRS and NFPA survey.

Table B-4. Smoking vs. Other Leading Major Fire Causes
by Selected Fire or Victim Characteristics
1994-1998 Home Smoking-Related Structure Fire Deaths
Reported to U.S. Fire Departments

A. Civilian Deaths

Major Cause of Fire	All Deaths	Condition = Asleep	Ignition Factor = Falling Asleep
Smoking materials	23.0%	27.2%	76.3%
Intentional	16.3%	8.9%	NA
Heating equipment	13.2%	16.8%	1.2%
Electrical distribution equipment	10.0%	12.2%	0.5%
Cooking equipment	9.3%	8.4%	12.8%
Child playing with fire	8.3%	6.2%	NA

NA: Not Applicable because "falling asleep" is identified using the Ignition Factor data element, which also is used to identify Intentional and Child Playing. Those two major causes cannot be entered when "falling asleep" is coded as the reason for the fire.

Note: These are national estimates of fires reported to U.S. municipal fire departments. Fires reported only to Federal or State agencies or industrial fire brigades are excluded. National estimates are projections. Casualty and loss projections can be influenced heavily by the inclusion or exclusion of one unusually serious fire. Statistics include a proportional allocation of fires where the major cause was unknown.

Source: National estimates based on NFIRS and NFPA survey.

Table B-5. Victim Location at Ignition vs. Victim Condition Before Injury, 1994-1998 Smoking-Material Home Structure Fires Reported to U.S. Fire Departments

A. Civilian Deaths

Location at Ignition	Asleep		Impaired		Disabled		Age-Limited		Awake		Total	
	% by Location	% by Condition	% by Location	% by Condition	% by Location	% by Condition	% by Location	% by Condition	% by Location	% by Condition	% by Location	% by Condition
Intimate with ignition	22.5%	44.9%	38.7%	20.0%	49.4%	24.0%	23.6%	2.5%	25.5%	5.8%	29.1%	100.0%
Not intimate, in room of origin	26.6%	55.3%	29.4%	15.9%	33.8%	17.1%	30.9%	3.5%	26.4%	6.2%	28.0%	100.0%
Not in room, on floor of origin	32.6%	67.0%	25.1%	13.4%	9.6%	4.8%	31.8%	3.5%	26.9%	6.3%	28.2%	100.0%
Not on floor, in building of origin	17.0%	76.0%	5.9%	6.8%	4.6%	5.0%	13.6%	3.3%	17.5%	8.9%	13.0%	100.0%
Outside of building of origin	0.4%	35.0%	0.0%	0.0%	1.7%	32.7%	0.0%	0.0%	1.7%	15.7%	0.7%	100.0%
Unclassified	0.8%	47.4%	0.9%	13.9%	1.0%	14.2%	0.0%	0.0%	2.0%	13.7%	1.0%	100.0%
Total	100.0%	58.0%	100.0%	15.1%	100.0%	14.2%	100.0%	3.1%	100.0%	6.6%	100.0%	100.0%

Note: These are national estimates of fires reported to U.S. municipal fire departments. Fires reported only to Federal or State agencies or industrial fire brigades are excluded. National estimates are projections. Casualty and loss projections can be influenced heavily by the inclusion or exclusion of one unusually serious fire. Statistics include a proportional allocation of fires where the victim location was unknown, for each condition before injury, and a proportional allocation of fires where the victim condition before injury was unknown. Shares for condition unclassified or under restraint are not shown. Totals may not equal sums because of rounding.

Source: National estimates based on NFIRS and NFPA survey.

Table B-5. Victim Location at Ignition vs. Victim Condition Before Injury, 1994-1998 Smoking-Material Home Structure Fires Reported to U.S. Fire Departments

B. Civilian Injuries

Location at Ignition	Asleep		Impaired		Disabled		Age-Limited		Awake		Total	
	% by Location	% by Condition	% by Location	% by Condition	% by Location	% by Condition	% by Location	% by Condition	% by Location	% by Condition	% by Location	% by Condition
Intimate with ignition	21.0%	52.6%	40.1%	17.3%	46.6%	9.6%	15.8%	1.8%	11.5%	17.3%	20.7%	100.0%
Not intimate, in room of origin	27.7%	56.0%	34.7%	12.1%	23.8%	4.0%	28.3%	2.6%	19.9%	24.1%	25.6%	100.0%
Not in room, on floor of origin	23.6%	55.6%	17.3%	7.0%	18.1%	3.5%	25.9%	2.7%	21.3%	30.0%	22.0%	100.0%
Not on floor, in building of origin	25.6%	54.8%	4.6%	1.7%	8.8%	1.5%	29.0%	2.8%	30.0%	38.5%	24.2%	100.0%
Outside of building of origin	1.0%	9.0%	2.7%	4.2%	1.4%	1.0%	0.0%	0.0%	14.6%	79.6%	5.7%	100.0%
Unclassified	1.1%	33.2%	0.6%	3.3%	1.3%	3.1%	1.0%	1.3%	2.7%	47.2%	1.8%	100.0%
Total	100.0%	51.9%	100.0%	8.9%	100.0%	4.3%	100.0%	2.3%	100.0%	31.0%	100.0%	100.0%

Note: These are national estimates of fires reported to U.S. municipal fire departments. Fires reported only to Federal or State agencies or industrial fire brigades are excluded. National estimates are projections. Casualty and loss projections can be influenced heavily by the inclusion or exclusion of one unusually serious fire. Statistics include a proportional allocation of fires where the victim location was unknown, for each condition before injury, and a proportional allocation of fires where the victim condition before injury was unknown. Shares for condition unclassified or under restraint are not shown. Totals may not equal sums because of rounding.

Source: National estimates based on NFIRS and NFPA survey.

Table B-6. Victim Condition Before Injury in Home Fire Casualties, Smoking-Related vs. Other Causes vs. All Causes 1994-1998 Structure Fires Reported to U.S. Fire Departments

A. Civilian Deaths

Condition Before Injury	Smoking	Other Cause	All Causes
Asleep	58.1%	50.3%	52.4%
Awake and unimpaired	6.6%	19.6%	16.1%
Impaired by alcohol or other drugs	15.1%	7.0%	9.2%
Physical handicap (including bedridden)	12.7%	6.9%	8.5%
Too young to act	0.9%	7.6%	5.8%
Too old to act	2.3%	3.1%	2.9%
Mental handicap (including senile)	1.4%	2.1%	1.9%
Under restraint	0.3%	0.2%	0.2%
Unclassified	2.7%	3.2%	3.0%
Total	100.0%	100.0%	100.0%

B. Civilian Injuries

Condition Before Injury	Smoking	Other Cause	All Causes
Awake and unimpaired	30.7%	59.1%	55.6%
Asleep	51.9%	30.3%	32.9%
Impaired by alcohol or other drugs	9.1%	1.9%	2.8%
Too young to act	0.8%	2.9%	2.7%
Physical handicap (including bedridden)	3.0%	1.5%	1.6%
Too old to act	1.5%	0.9%	1.0%
Mental handicap (including senile)	1.4%	0.8%	0.9%
Under restraint	0.1%	0.1%	0.1%
Unclassified	1.6%	2.5%	2.4%
Total	100.0%	100.0%	100.0%

Note: These are national estimates of fires reported to U.S. municipal fire departments. Fires reported only to Federal or State agencies or industrial fire brigades are excluded. National estimates are projections. Casualty and loss projections can be influenced heavily by the inclusion or exclusion of one unusually serious fire. Statistics include a proportional allocation of fires where the victim condition was unknown.

Source: National estimates based on NFIRS and NFPA survey.

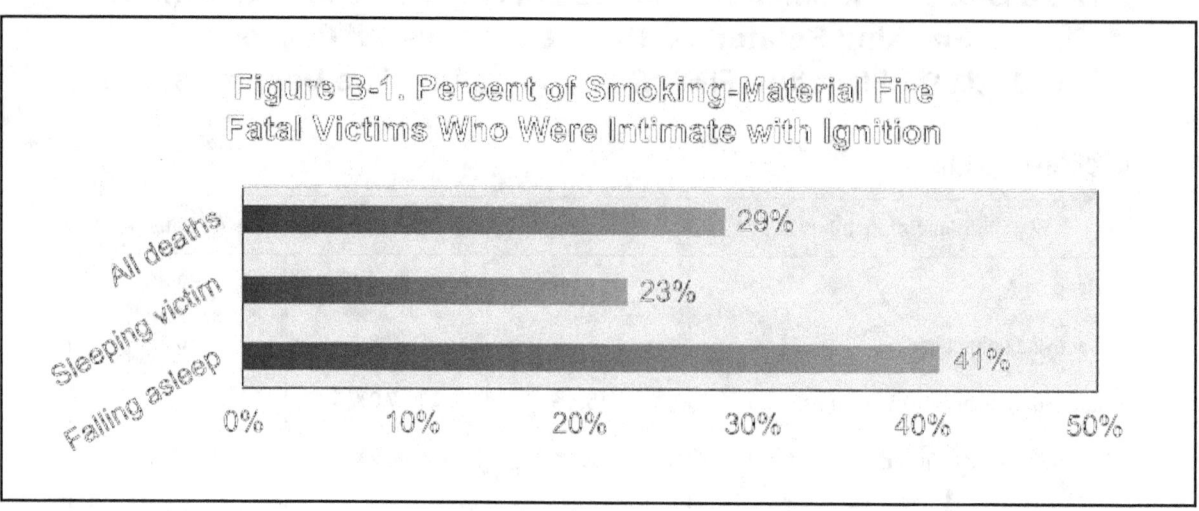

Figure B-1. Percent of Smoking-Material Fire Fatal Victims Who Were Intimate with Ignition

- All deaths: 29%
- Sleeping victim: 23%
- Falling asleep: 41%

Note: These are national estimates of fires reported to U.S. municipal fire departments. Fires reported only to Federal or State agencies or industrial fire brigades are excluded. National estimates are projections. Casualty and loss projections can be influenced heavily by the inclusion or exclusion of one unusually serious fire. Statistics include a proportional allocation of fires where the victim location was unknown. Sleeping victim is based on Condition Before Injury. Falling asleep is based on Ignition Factor and refers to circumstances regarding the cause of the fire, not the victim.

Source: National estimates based on NFIRS and NFPA survey.

Table B-7. Leading Items First Ignited by Victim Location at Ignition
Percent of 1994-1998 Smoking-Material Home Structure Fire Deaths
Reported to U.S. Fire Departments

Item First Ignited	All Victims	Victims Intimate with Ignition
Upholstered furniture	46.7%	34.2%
Mattress or bedding	27.1%	35.6%
Clothing	6.5%	15.0%
Trash	3.1%	1.8%
Multiple items first ignited	2.6%	1.4%
Unknown-type furniture	2.1%	1.0%
Papers	1.6%	1.1%
Interior wall covering	1.5%	1.2%
Structural member or framing	1.4%	0.8%
Unclassified item	1.1%	1.2%
Unknown-type soft goods or clothing	0.9%	1.7%
Unclassified furniture	0.7%	0.0%

Note: These are national estimates of fires reported to U.S. municipal fire departments. Fires reported only to Federal or State agencies or industrial fire brigades are excluded. National estimates are projections. Casualty and loss projections can be influenced heavily by the inclusion or exclusion of one unusually serious fire. Statistics include a proportional allocation of smoking-material fires where the form of material first ignited was unknown.

Source: National estimates based on NFIRS and NFPA survey.

Table B-8. Trend in Share of Certain Leading Materials First Ignited in Home Structure Fire Deaths, 1980-1998 Structure Fires Reported to U.S. Fire Departments

Year	Percent Share Involving Mattress, Bedding, or Upholstered Furniture
1980	85%
1981	83%
1982	86%
1983	86%
1984	87%
1985	80%
1986	80%
1987	76%
1988	81%
1989	85%
1990	79%
1991	84%
1992	78%
1993	82%
1994	72%
1995	73%
1996	73%
1997	75%
1998	77%

Note: These are national estimates of fires reported to U.S. municipal fire departments. Fires reported only to Federal or State agencies or industrial fire brigades are excluded. National estimates are projections. Casualty and loss projections can be influenced heavily by the inclusion or exclusion of one unusually serious fire. Statistics include a proportional allocation of smoking-material fires where the form of material first ignited was unknown.

Source: National estimates based on NFIRS and NFPA survey.

Table B-9. Victim Location at Ignition vs. Activity When Injured, Smoking-Related vs. Other Causes vs. All Causes 1994-1998 Home Structure Fires Reported to U.S. Fire Departments

A. Civilian Deaths

Location at Ignition	Sleeping			Attempting to Escape			Acting Irrationally or Unable to Act			Attempting to Fight Fire or Attempting Rescue		
	Smoking	Other Cause	All Causes	Smoking	Other Cause	All Causes	Smoking	Other Cause	All Causes	Smoking	Other Cause	All Causes
Intimate with ignition	29.9%	2.7%	11.2%	19.5%	6.4%	9.6%	44.7%	26.0%	30.6%	17.6%	14.6%	15.4%
Not intimate, in room of origin	26.3%	18.6%	21.0%	21.7%	19.3%	19.8%	30.5%	29.8%	30.0%	35.2%	25.9%	28.3%
Not in room, on floor of origin	30.5%	41.2%	37.8%	33.2%	36.9%	36.0%	17.6%	22.5%	21.3%	30.7%	24.2%	25.9%
Not on floor, in building of origin	12.4%	36.6%	29.1%	24.5%	36.6%	33.7%	5.4%	19.0%	15.6%	11.9%	27.2%	23.3%
Outside of building of origin	0.9%	0.5%	0.6%	0.0%	0.8%	0.6%	0.5%	0.6%	0.6%	2.3%	6.8%	5.7%
Unclassified	0.0%	0.5%	0.3%	1.0%	0.0%	0.2%	1.4%	2.0%	1.9%	2.3%	1.2%	1.5%
Total	100.0%	100.0%	100.0%	100.0%	100.0%	100.0%	100.0%	100.0%	100.0%	100.0%	100.0%	100.0%

Note: These are national estimates of fires reported to U.S. municipal fire departments. Fires reported only to Federal or State agencies or industrial fire brigades are excluded. National estimates are projections. Casualty and loss projections can be influenced heavily by the inclusion or exclusion of one unusually serious fire. Statistics include a proportional allocation of fires where the victim location was unknown.

Source: National estimates based on NFIRS and NFPA survey.

Table B-9. Victim Location at Ignition vs. Activity When Injured, Smoking-Related vs. Other Causes vs. All Causes 1994-1998 Home Structure Fires Reported to U.S. Fire Departments (Cont'd)

B. Civilian Injuries

Location at Ignition	Sleeping			Attempting to Escape			Acting Irrationally or Unable to Act			Attempting to Fight Fire or Attempting Rescue		
	Smoking	Other Cause	All Causes	Smoking	Other Cause	All Causes	Smoking	Other Cause	All Causes	Smoking	Other Cause	All Causes
Intimate with ignition	31.3%	4.2%	10.4%	7.8%	7.2%	7.2%	39.8%	33.6%	34.4%	14.4%	13.6%	13.7%
Not intimate, in room of origin	29.3%	21.8%	23.5%	17.6%	17.7%	17.7%	28.5%	26.9%	27.2%	27.2%	33.2%	32.7%
Not in room, on floor of origin	19.3%	38.3%	34.0%	34.7%	31.3%	31.7%	15.3%	18.7%	18.3%	21.7%	20.8%	20.9%
Not on floor, in building of origin	18.5%	33.1%	29.8%	36.5%	39.8%	39.4%	12.6%	14.9%	14.6%	18.6%	17.8%	17.8%
Outside of building of origin	0.6%	1.3%	1.2%	2.3%	2.0%	2.1%	2.9%	3.8%	3.7%	15.5%	12.2%	12.5%
Unclassified	1.0%	1.2%	1.2%	1.1%	2.0%	1.9%	0.9%	2.1%	1.9%	2.6%	2.4%	2.5%
Total	100.0%	100.0%	100.0%	100.0%	100.0%	100.0%	100.0%	100.0%	100.0%	100.0%	100.0%	100.0%

Note: These are national estimates of fires reported to U.S. municipal fire departments. Fires reported only to Federal or State agencies or industrial fire brigades are excluded. National estimates are projections. Casualty and loss projections can be influenced heavily by the inclusion or exclusion of one unusually serious fire. Statistics include a proportional allocation of fires where the victim location was unknown.

Source: National estimates based on NFIRS and NFPA survey.

Table B-10. Victim Condition Before Injury vs. Activity When Injured, Smoking-Related vs. Other Causes vs. All Causes 1994-1998 Home Structure Fires Reported to U.S. Fire Departments

A. Civilian Deaths

Condition Before Injury	Sleeping			Attempting to Escape			Acting Irrationally or Unable to Act			Attempting to Fight Fire or Attempting Rescue		
	Smoking	Other Cause	All Causes	Smoking	Other Cause	All Causes	Smoking	Other Cause	All Causes	Smoking	Other Cause	All Causes
Asleep	82.3%	89.4%	87.3%	65.8%	57.7%	59.6%	16.6%	10.9%	12.3%	11.3%	34.1%	28.4%
Awake and unimpaired	0.2%	2.3%	1.7%	9.5%	21.8%	18.8%	4.7%	25.7%	20.5%	59.9%	49.9%	52.5%
Impaired by alcohol or other drugs	12.3%	3.4%	6.2%	8.8%	4.0%	5.2%	24.9%	10.3%	13.9%	13.4%	4.0%	6.4%
Physical handicap (including bedridden)	1.8%	2.3%	2.1%	9.8%	5.0%	6.1%	40.9%	11.4%	18.7%	6.3%	4.1%	4.7%
Too young to act	0.6%	2.0%	1.6%	0.0%	6.3%	4.8%	0.9%	26.1%	19.8%	0.0%	0.8%	0.6%
Too old to act	0.3%	0.6%	0.5%	2.5%	2.3%	2.4%	5.8%	5.4%	5.5%	0.0%	4.3%	3.2%
Mental handicap (including senile)	0.8%	0.0%	0.2%	1.4%	1.1%	1.1%	0.8%	6.3%	5.0%	2.8%	0.0%	0.6%
Under restraint	0.6%	0.0%	0.1%	0.0%	0.0%	0.0%	0.0%	1.0%	0.8%	0.0%	0.0%	0.0%
Unclassified	1.1%	0.0%	0.3%	2.2%	1.8%	1.9%	5.5%	3.0%	3.6%	6.3%	2.8%	3.7%
Total	100.0%	100.0%	100.0%	100.0%	100.0%	100.0%	100.0%	100.0%	100.0%	100.0%	100.0%	100.0%

Note: These are national estimates of fires reported to U.S. municipal fire departments. Fires reported only to Federal or State agencies or industrial fire brigades are excluded. National estimates are projections. Casualty and loss projections can be influenced heavily by the inclusion or exclusion of one unusually serious fire. Statistics include a proportional allocation of fires where the victim condition was unknown.

Source: National estimates based on NFIRS and NFPA survey.

Table B-10. Victim Condition Before Injury vs. Activity When Injured, Smoking-Related vs. Other Causes vs. All Causes 1994-1998 Home Structure Fires Reported to U.S. Fire Departments (Cont'd)

B. Civilian Injuries

Condition Before Injury	Sleeping			Attempting to Escape			Acting Irrationally or Unable to Act			Attempting to Fight Fire or Attempting Rescue		
	Smoking	Other Cause	All Causes	Smoking	Other Cause	All Causes	Smoking	Other Cause	All Causes	Smoking	Other Cause	All Causes
Awake and unimpaired	1.5%	2.3%	2.1%	28.3%	44.9%	42.9%	25.6%	54.6%	50.7%	61.1%	82.2%	80.3%
Asleep	89.1%	93.4%	92.5%	61.2%	46.6%	48.3%	12.6%	5.9%	6.8%	28.3%	13.0%	14.4%
Impaired by alcohol or other drugs	7.6%	1.7%	3.0%	4.8%	1.0%	1.5%	26.8%	7.5%	10.1%	5.0%	1.0%	1.3%
Too young to act	0.2%	1.0%	0.8%	0.8%	2.6%	2.4%	4.5%	15.9%	14.4%	0.0%	0.1%	0.1%
Physical handicap (including bedridden)	1.0%	1.0%	1.0%	2.5%	1.4%	1.5%	16.1%	6.1%	7.4%	0.6%	0.4%	0.4%
Too old to act	0.4%	0.2%	0.2%	1.4%	1.3%	1.3%	4.6%	3.6%	3.8%	0.8%	0.2%	0.3%
Mental handicap (including senile)	0.1%	0.1%	0.1%	0.8%	0.6%	0.6%	7.4%	3.7%	4.2%	0.7%	0.3%	0.3%
Under restraint	0.1%	0.0%	0.0%	0.0%	0.1%	0.1%	0.0%	0.2%	0.2%	0.0%	0.1%	0.1%
Unclassified	0.1%	0.3%	0.2%	0.2%	1.5%	1.3%	2.3%	2.4%	2.4%	3.5%	2.7%	2.8%
Total	100.0%	100.0%	100.0%	100.0%	100.0%	100.0%	100.0%	100.0%	100.0%	100.0%	100.0%	100.0%

Note: These are national estimates of fires reported to U.S. municipal fire departments. Fires reported only to Federal or State agencies or industrial fire brigades are excluded. National estimates are projections. Casualty and loss projections can be influenced heavily by the inclusion or exclusion of one unusually serious fire. Statistics include a proportional allocation of fires where the victim condition was unknown.

Source: National estimates based on NFIRS and NFPA survey.

Table B-11. Condition Preventing Victim Escape in Home Fire Casualties, Smoking-Related vs. Other Causes vs. All Causes 1994-1998 Structure Fires Reported to U.S. Fire Departments

A. Civilian Deaths

Condition Preventing Escape	Smoking	Other Cause	All Causes
Insufficient escape time or rapid fire progress	15.6%	27.6%	24.4%
Fire between person and exit	17.2%	22.2%	20.9%
No condition prevented escape	14.2%	16.0%	15.5%
Incapacitated before ignition	24.5%	10.5%	14.3%
Moved too slowly or failed to follow escape procedures	11.2%	8.1%	8.9%
Clothing on person burning	8.8%	4.7%	5.8%
Locked door	2.1%	3.0%	2.7%
Illegal gate or lock	0.3%	0.2%	0.2%
Unclassified	6.2%	7.8%	7.4%
Total	**100.0%**	**100.0%**	**100.0%**

B. Civilian Injuries

Location at Ignition	Smoking	Other Cause	All Causes
No condition prevented escape	59.7%	68.3%	67.2%
Insufficient escape time or rapid fire progress	6.9%	11.5%	10.9%
Fire between person and exit	9.3%	6.6%	6.9%
Moved too slowly or failed to follow escape procedures	6.9%	3.4%	3.8%
Clothing on person burning	2.3%	3.7%	3.5%
Incapacitated before ignition	9.2%	2.3%	3.2%
Locked door	1.1%	0.6%	0.7%
Illegal gate or lock	0.2%	0.1%	0.1%
Unclassified	4.5%	3.5%	3.7%
Total	**100.0%**	**100.0%**	**100.0%**

Note: These are national estimates of fires reported to U.S. municipal fire departments. Fires reported only to Federal or State agencies or industrial fire brigades are excluded. National estimates are projections. Casualty and loss projections can be influenced heavily by the inclusion or exclusion of one unusually serious fire. Statistics include a proportional share of fires with an unknown condition preventing escape.

Source: National estimates based on NFIRS and NFPA survey.

Table B-12. Location of Fire vs. Location of Fatal Victim, by Whether Victim Was the Smoker Whose Cigarette Started the Fire FIDO Fatal Smoking-Related Home Fires from 1997-1998

A. Full Display

Victim Location When Found	Area of Fire Origin	Total Deaths	Smoker	Not the Smoker	Unknown Whether the Smoker
Bathroom	Bedroom	13	12	0	1
Bathroom	Living room, family room, or den	5	5	0	0
Bathroom	Other	1	0	0	1
Bedroom	Bedroom	64	50	5*	9
Bedroom	Kitchen	6	2	0	4
Bedroom	Living room, family room, or den	59	21	28	10
Bedroom	Other	5	1	2	2
Garage	Other	2	2	0	0
Kitchen	Bedroom	4	4	0	0
Kitchen	Kitchen	2	1	0	1
Kitchen	Living room, family room, or den	12	10	2	0
Kitchen	Other	3	2	1	0
Kitchen	Yard	1	0	0	1
Living room, family room, or den	Bedroom	8	6	1	1
Living room, family room, or den	Kitchen	4	3	0	1
Living room, family room, or den	Living room, family room, or den	63	51	6	6
Living room, family room, or den	Other	1	1	0	0
Porch or balcony	Living room, family room, or den	3	3	0	0
Yard	Bedroom	6	4	0	2
Yard	Living room, family room, or den	2	1	0	1
Other	Bedroom	10	6	2	2
Other	Living room, family room, or den	18	12	6	0
Other	Other	7	4	1	2
Unknown	Bedroom	21	15	4	2
Unknown	Kitchen	3	0	3	0
Unknown	Living room, family room, or den	53	26	16	11
Unknown	Other, specify:	13	2	0	11
Total	Total	389	244	77	68
Percent of known total		100%	76%	24%	--

*Only 1 of these 5 incidents involved the same bedroom for victim location and area of fire origin.

Source: FIDO analysis for smoker behavior mitigation project.

Table B-12. Location of Fire vs. Location of Fatal Victim, by Whether Victim Was the Smoker Whose Cigarette Started the Fire FIDO Fatal Smoking-Related Home Fires from 1997-1998 (Cont'd)

B. Totals for Victim Location When Found

Victim Location When Found	Area of Fire Origin	Total Deaths	Smoker	Not the Smoker	Unknown Whether the Smoker
Bathroom	All	19	17	0	2
Bedroom	All	134	74	35	25
Garage	All	2	2	0	0
Kitchen	All	22	17	3	2
Living room, family room, or den	All	76	61	7	8
Porch or balcony	All	3	3	0	0
Yard	All	8	5	0	3
Other	All	35	22	9	4
Unknown	All	90	43	23	24
Total	Total	389	244	77	68

Source: FIDO analysis for smoker behavior mitigation project.

C. Totals for Area of Fire Origin

Victim Location When Found	Area of Fire Origin	Total Deaths	Smoker	Not the Smoker	Unknown Whether the Smoker
All	Bedroom	126	97	12	17
All	Kitchen	15	6	3	6
All	Living room, family room, or den	215	129	58	28
All	Yard	1	0	0	1
All	Other	32	12	4	16
Total	Total	389	244	77	68

Source: FIDO analysis for smoker behavior mitigation project.

Table B-13. Location of Fire vs. Location of Fatal Victim, by Victim Relationship to Smoker Whose Cigarette Started the Fire
FIDO Fatal Smoking-Related Home Fires from 1997-1998

Victim Location When Found	Area of Fire Origin	Total Not Smoker	Child of Smoker	Neighbor or Friend of Smoker	Spouse or Partner of Smoker	Parent of Smoker	Other
Bedroom	Bedroom	5	1	0	2*	2	0
Bedroom	Living room, family room, or den	28	16	3	2	3	4
Bedroom	Other	2	2	0	0	0	0
Kitchen	Living room, family room, or den	2	0	0	2	0	0
Kitchen	Other	1	0	1	0	0	0
Living room, family room, or den	Bedroom	1	0	0	0	0	1
Living room, family room, or den	Living room, family room, or den	6	3	2	1	0	0
Other	Bedroom	2	0	2	0	0	0
Other	Living room, family room, or den	6	0	2	2	2	0
Other	Other	1	0	0	0	0	1
Unknown	Bedroom	4	1	2	0	0	1
Unknown	Kitchen	3	0	3	0	0	0
Unknown	Living room, family room, or den	16	3	4	2	3	4
Total		77	26	19	11	10	11
Percent of known total		100%	34%	25%	14%	13%	14%

*1 of these 2 is the only Bedroom/Bedroom combination involving the same bedroom.

Note: Some smoker's children were adults. Most neighbors were on site, like friends, before fire began.

Source: FIDO analysis for smoker behavior mitigation project.

Table B-14. Victim Condition, by Whether Fatal Victim Was the Smoker Whose Cigarette Started the Fire
FIDO Fatal Smoking-Related Home Fires from 1997-1998

A. Full Display

Limitation	Limitation	Limitation	Total Deaths	Smoker	Not the Smoker	Unknown Whether the Smoker
Alcohol			35	26	3	6
Alcohol	Drugs		4	4	0	0
Alcohol	Drugs	Physical disability--not age	1	1	0	0
Alcohol	Drugs	Sleepy	6	3	0	3
Alcohol	Mental disability--not age		1	1	0	0
Alcohol	Other		8	6	0	2
Alcohol	Oxygen		1	1	0	0
Alcohol	Oxygen	Sleepy	1	1	0	0
Alcohol	Physical disability--not age		3	3	0	0
Alcohol	Physical disability--not age	Sleepy	8	8	0	0
Alcohol	Physical limitation--age		1	1	0	0
Alcohol	Physical limitation--age	Sleepy	2	1	1	0
Alcohol	Sleepy		60	50	4	6
Alcohol	Sleepy	Other	1	1	0	0
Drugs			1	1	0	0
Drugs	Physical disability--not age		2	2	0	0
Drugs	Physical disability--not age	Sleepy	1	1	0	0
Drugs	Physical limitation--age	Sleepy	1	1	0	0
Drugs	Sleepy		4	2	0	2
Mental disability--not age	Physical limitation--age		1	1	0	0
Mental disability- -not age	Physical limitation--age	Sleepy	1	1	0	0
Mental disability-not age	Sleepy		3	2	1	0
Mental limitation--age	Physical limitation--age		2	1	1	0
Mental limitation--age	Physical limitation--age	Sleepy	3	2	1	0

Table B-14. Victim Condition, by Whether Fatal Victim Was the Smoker Whose Cigarette Started the Fire FIDO Fatal Smoking-Related Home Fires from 1997-1998 (Cont'd)

A. Full Display

Limitation	Limitation	Limitation	Total Deaths	Smoker	Not the Smoker	Unknown Whether the Smoker
Oxygen			2	2	0	0
Oxygen	Other		1	1	0	0
Oxygen	Physical disability--not age		4	4	0	0
Oxygen	Physical limitation--age		4	4	0	0
Oxygen	Physical limitation--age	Sleepy	3	3	0	0
Oxygen	Sleepy		1	1	0	0
Physical disability--not age			9	7	0	2
Physical disability--not age	Other		2	2	0	0
Physical disability--not age	Physical limitation--age		1	0	1	0
Physical disability--not age	Sleepy		14	8	4	2
Physical limitation--age			16	9	5	2
Physical limitation--age	Sleepy		25	12	9	4
Sleepy			96	43	33	20
Sleepy	Other		1	1	0	0
Other			6	5	0	1
None			53	21	14	18
Total			**389**	**244**	**77**	**68**
Percent of known total			**100%**	**76%**	**24%**	**--**

Source: FIDO analysis for smoker behavior mitigation project.

Table B-14. Victim Condition, by Whether Fatal Victim Was the Smoker Whose Cigarette Started the Fire
FIDO Fatal Smoking-Related Home Fires from 1997-1998 (Cont'd)

B. Totals for Limitations

Limitation	Total Deaths	Smoker	Not the Smoker	Unknown Whether the Smoker
Alcohol	132	107	8	17
Drugs	20	15	0	5
Mental disability--not age-related	6	5	1	0
Mental limitation--age-related	5	3	2	0
Oxygen (medical) in use	17	17	0	0
Physical disability--not age-related	45	36	5	4
Physical limitation--age-related	60	36	18	6
Sleepy	231	141	53	37
Other	19	16	0	3
None	53	21	14	18

Source: FIDO analysis for smoker behavior mitigation project.

C. Totals for Limitations with No Other Limitations

Limitation	Total Deaths	Smoker	Not the Smoker	Unknown Whether the Smoker
Alcohol	35	26	3	6
Drugs	1	1	0	0
Mental disability--not age-related	0	0	0	0
Mental limitation--age-related	0	0	0	0
Oxygen (medical) in use	2	2	0	0
Physical disability--not age-related	9	7	0	2
Physical limitation--age-related	16	9	5	2
Sleepy	96	43	33	20
Other	6	5	0	1
None	53	21	14	18

Source: FIDO analysis for smoker behavior mitigation project.

Table B-15. Victim Condition, by Fatal Victim Relationship to Smoker Whose Cigarette Started the Fire FIDO Fatal Smoking-Related Home Fires from 1997-1998

Limitation	Limitation	Limitation	Total Not Smoker	Child of Smoker	Neighbor or Friend of Smoker	Spouse or Partner of Smoker	Parent of Smoker	Other
Alcohol			3	0	1	1	0	1
Alcohol	Physical limitation--age	Sleepy	1	0	1	0	0	0
Alcohol	Sleepy		4	0	2	0	0	2
Mental disability--not age	Sleepy		1	0	0	0	0	1
Mental limitation--age			1	1	0	0	0	0
Mental limitation--age	Physical limitation--age	Sleepy	1	1	0	0	0	0
Physical disability--not age			1	0	0	0	1	0
Physical disability--not age	Sleepy		4	0	0	2	2	0
Physical limitation--age			5	1	3	1	0	0
Physical limitation--age	Sleepy		9	3	3	0	3	0
Sleepy			33	16	4	4	2	7
None			14	4	5	3	2	0
Total			77	26	19	11	10	11
Percent of known total			100%	34%	25%	14%	13%	14%

Note: Some smoker's children were adults. Most neighbors were on site, like friends, before fire began.

Source: FIDO analysis for smoker behavior mitigation project.

Appendix C

Additional Information on Smoking and Medical Oxygen

Philadelphia's Deputy Chief Garrity described a public education program to prevent oxygen therapy fires in his 2000 paper for the Executive Fire Officer Program at the National Fire Academy (NFA). From March 3, 1999, to November 30, 2000, 12 oxygen-therapy fires in Philadelphia caused 3 deaths and injured 7 others. Although the cause of six fires was listed as open-flame, in all six cases, the open flame was being used to light a cigarette. The remaining six were specifically caused by smoking.

He noted that long-term smokers find it difficult to quit even in the face of requirements that patients not smoke. Their public education program included a message not to smoke **while using oxygen** and instructs caregivers to remove all smoking materials and implements from the oxygen user's room. Their program included a television public service announcement and a school supplement on the topic to enlist the support of children and grandchildren.[1]

In a November 2003 safety brochure, the Massachusetts Office of the State Fire Marshal reported that "Since 1997, 16 people have died and 20 other individuals have suffered severe burns or smoke inhalation in fires involving people who were smoking while using home oxygen systems." People are advised never to smoke or light a match while using oxygen, to keep all heat sources away from oxygen equipment, and not to allow smoking inside a home where oxygen is used. Even if the oxygen has been shut off, the environment still may be oxygen-enriched.[2]

In a telephone conversation with Jennifer Mieth, Public Information Officer, on March 10, 2004, she commented that they were considering telling smokers to shut off the oxygen, wait 10 minutes, and then go outside to smoke. In a recent fatal fire, a clothing ignition occurred when an oxygen user lit up after shutting off the flow of oxygen. She theorized that the victim's clothing was still oxygen-saturated.

The Joint Commission on Accreditation of Healthcare Organizations (JCAHO) identified a number of risk factors for fires occurring when home health care patients were using supplemental oxygen. These included living alone, the absence of working smoke alarms, cognitive impairment, a history of smoking while oxygen was in use, and flammable clothing. Recommendations included better staff training, improved communication among providers; involving ethics committees in decisions to end services to noncompliant patients; and increasing fire safety with smoke alarms and other practices.[3]

Alisa Wolf's 1998 article in *NFPA Journal* describes the growing number of people who require medical procedures and equipment while still living in the community. The JCAHO has standards for home health care organizations, but accreditation is not mandatory for home health care Medicare participation. The home environment is much less tightly regulated than hospitals or nursing homes. People may not comply with recommendations. They may continue to smoke despite the danger. One fire chief reported a patient releasing oxygen to cool a room.[4]

References

1. Garrity, Thomas J. *A Public Education Program to Prevent Oxygen-Therapy Fires: Strategic Analysis of Community Risk Reduction.* National Fire Academy applied research project as part of the Executive Fire Officer Program, Emmitsburg, MD, 2000.

2. Office of the State Fire Marshal, Massachusetts Department of Fire Services. "Smoking and Home Oxygen Systems: 'Some People Don't Know When to Quit.'" (Nov. 2003), online at http://www.state. ma.us/dfs/osfm/pubed/flyers/Smoking_and_ Home_Oxygen.pdf

3. Joint Commission on Accreditation of Healthcare Organizations. "Lessons Learned: Fires in the Home Care Setting." *Sentinel Event Alert* 17, (Mar. 2001), online at http://www.jcaho.org/about+us/news+letters/ sentinel+event+alert/print/sea_17.htm

4. Wolf, Alisa. "When Health Care Moves Home." *NFPA Journal*, 92, no. 1 (1998): 64-67.

Appendix D

Additional Information on Reduced-Ignition-Strength Cigarettes

Barillo et al. state, "Because many smoking-related fire fatalities involve alcohol use, modification of human behavior is unlikely to be successful."[1] The first Federal legislation on the topic of fire-safe (or, more accurately, reduced-ignition-strength) cigarettes was introduced in the late 1920's as a tool to prevent forest fires.[1]

In a 1933 article, Hoffheins described tests done to determine if modified cigarettes would reduce the likelihood of grass or forest floor ignition. They found that the addition of cigarette tips made of cigarette paper with less inorganic material reduced the number of cigarette ignitions and did not displease the consumer.[2]

In their study comparing smokers who did and did not have fires and the brands of cigarettes smoked by each group, Karter et al. found that cigarettes involved in fires were less likely to have filters, tended to have a shorter filter if a filter was present, and tended to have greater porosity of cigarette wrapping paper. The type of pack was relevant for males only; the risk of fire was much higher if a man smoked a cigarette from a soft pack rather than a hard pack.[3]

If any of the experimental cigarettes evaluated by the Technical Study Group under the Cigarette Safety Act of 1984 had been in exclusive use, the smoking-material fire death toll would have been projected to decrease by 58 percent in 1986, and it was expected that these deaths would be reduced by 64 percent by 1996.[4] In 2001, Gann et al. reported that banded cigarettes test-marketed by a major manufacturer had a lower ignition propensity in the test scenarios than did conventional cigarettes.[5]

Effective June 28, 2004, cigarettes sold in the State of New York must have a low ignition strength and be more likely to self-extinguish if unattended.[6]

Outside the U.S., on March 31, 2004, the Canadian Parliament passed Bill C-260. This bill requires cigarettes to self-extinguish if left unsmoked. Provisions of the bill went into effect at the end of 2004.[7]

On April 2, 2004, Representative Edward J. Markey and Representative Peter King introduced the Cigarette Fire Safety Act of 2004. The bill reiterates U.S. statistics about cigarette fire losses, and notes that, because of the passage of two bills promoted by Joseph Moakley, the Cigarette Safety Act of 1984 and the Fire Safe Cigarette Act of 1990, the technical work for a standard has been done. The proposed bill would require the U.S. Consumer Product Safety Commission (CPSC) to prescribe at least one cigarette fire safety standard similar to New York's within 18 months of enactment.[8,9] The bill was not enacted.

There may be perceived adverse side effects to a reduced ignition strength cigarette. For example, smokers may anticipate a less enjoyable smoking experience or a higher cost. The tobacco industry has encouraged these perceptions in its testimony on bills to require such cigarettes. However, since the adoption of a requirement in New York, early evidence shows no higher cost (and no reason to expect costs to go higher) and no clear evidence of a decline in smoking or a massive attempt to circumvent the regulation in order to get and enjoy noncompliant cigarettes. At the same time, these are reactions to a legislative requirement. Smoker perceptions still appear to be a major barrier to any large-scale voluntary, market-driven move to reduced-ignition-strength cigarettes.

The safety community also must keep abreast of new products. Bidis, filterless cigarettes wrapped in a leaf, are usually imported from India. They often are flavored and are seen as an entry for teens into tobacco use. They are said to go out more easily than conventional cigarettes, and must be relit frequently.[10] New products may be outside the purview of existing legislation or regulations.

References

1. Barillo, David J., Peter A. Brigham, Debra Ann Kayden, Robert T. Heck, and Albert T. McManus. "The Fire Safe Cigarette: A Burn Prevention Tool." *Journal of Burn Care and Rehabilitation*, (Mar./Apr. 2000): 164-170.

2. Hoffheins, F.M. "Fire Hazard Tests with Cigarettes," *Quarterly of the National Fire Protection Association* 27, no. 2, (Oct. 1933): 132-140.

3. Karter, Michael J., Terry L. Kissinger, Alison L. Miller, Beatrice Harwood, Rita F. Fahy, and John R. Hall, Jr. "Cigarette Characteristics, Smoker Characteristics, and the Relationship to Cigarette Fires." *Fire Technology* 30, no. 4, (1994): 400-431.

4. Hall, John R., Jr. *et al.*, "Expected Changes in Fire Damages from Reducing Cigarette Ignition Propensity." vol. 5 of *Report of the Technical Study Group Cigarette Safety Act of 1984*, Washington: U.S. Consumer Product Safety Commission, 1987.

5. Gann, Richard, Kenneth D. Steckler, Schuyler Ruitberg, William F. Guthrie, and Mark S. Levenson. *Relative Ignition Propensity of Test Market Cigarettes*. National Institute of Standards and Technology, 2001, online at http://fire.nist.gov/bfrlpubs/fire01/PDF/f01007.pdf

6. Department of State, New York. "New York State Adopts Nation's First Cigarette Fire Safety Standard." Press Release Dec. 31, 2003, online at http://www.dos.state.ny.us/pres/pr2003/12_31.htm

7. Cherner, Joe (announced by). "Canada Becomes First Country to Require Fire-Safe Cigarettes," Mar. 31, 2004, online at http://www.smokefree.net/JoeCherner-announce/messages/247275.html

8. Markey, Edward. "Statement of Introduction: The Cigarette Fire Safety Act of 2004." Apr. 2, 2004, online at http://www.house.gov/markey/Issues/iss_cigfs_pr040402.pdf

9. Markey, Edward. "A Bill to Provide for Fire Safety Standards for Cigarettes, and for Other Purposes" Apr. 2, 2004, online at http://www.house.gov/markey/Issues/iss_cigfs_bill040402.pdf

10. Fisher, Laurie. "Bidis--The Latest Trend in U.S. Teen Tobacco Use." *Cancer Causes and Control*, Kluwer Academic Publishers, Printed in the Netherlands 11 (2000): 577-578, online at http://www.hsph.harvard.edu/cancer/research/tobacco/tobacco%20notes%20archive/tobacco_notes_11_577_578_2000

Appendix E

Implementation of Mitigation Strategies Into USFA Public Fire Safety Education Materials

The following seven educational messages were developed for this project:

- If you smoke, smoke outside.

- Wherever you smoke, use deep, sturdy ashtrays.

- Before you throw out butts and ashes, make sure they are out, and dowsing in water or sand is the best way to do that.

- Check under furniture cushions and in other places people smoke for cigarette butts that may have fallen out of sight.

- Smoking should not be allowed in a home where oxygen is used.

- If you smoke, choose fire-safe cigarettes. They are less likely to cause fires.

- To prevent a deadly cigarette fire, you have to be alert. You won't be if you are sleepy, have been drinking, or have taken medicine or other drugs.

These recommended messages should be applied to existing USFA educational materials, with prioritizing based on space available, as described below:

Protecting Your Family From Fire (FA-130/August 2002)

Changes to this short brochure probably would require a more general rethinking of the structure and space allocation priorities of the brochure as a whole. Currently, cause-related information is limited to a fraction of one paragraph. Roughly 3 pages and 11 paragraphs are devoted to causes, toxicity, and the special risks of the very young and very old.

If this brochure were restructured along cause lines-- with a paragraph each on safety tips--it could cover smoking, heating, cooking, candle (primary part of open flame), electrical distribution equipment, child-playing fires, and also upholstered furniture,

mattresses and bedding, and clothing fires. Special concerns of the very young or very old could be inserted in each cause paragraph, and there would be one paragraph each on just the high risks of these two groups to go with the nine paragraphs on nine causes.

With this approach, the one paragraph on smoking fires could have room for the first 4 bullets under recommended messages and might be able to also include the last bullet on alertness. Treat the recommendations as a prioritized list and include messages up to the space limits.

Is Your Home Fire Safe? Door Knob Hanger (FA-285/August 1999)

This includes three bullets under a headline of "Safe Smoking Habits." This would be more consistent with the other cause-related sections if the headline were changed to simply "Smoking." Given the severe space limits, include the first two bullets and all or the first half of the third bullet in the recommended messages.

Fire Safety Checklist for Older Adults (FA-221/August 2002)

Older adults are probably the most resistant to significant behavior change. They may be at heightened risk (e.g., hypothermia) if pressed to smoke outside. They also are particularly in need of the alertness and medical-oxygen messages. This paragraph incorporates the second, third, and fourth bullets and the last one on alertness:

If You Smoke, Avoid Fire. Wherever you smoke, use deep, sturdy ashtrays. Smoking should not be allowed in a home where oxygen is used. Before you throw out butts and ashes, make sure they are out, and dowsing in water or sand is the best way to do that. Check under furniture cushions and in other places people smoke for cigarette butts that may have fallen out of sight. To prevent a deadly cigarette fire, you have to be alert. You won't be if you are sleepy, have been drinking, or have taken medicine or other drugs.

Fire Risks for the Blind or Visually Impaired (FA-205, 12/99)
Fire Risks for the Deaf or Hard of Hearing (FA-202, 12/99)
Fire Risks for the Mobility Impaired (FA-204, 12/99)
Fire Risks for the Older Adult (FA-203, 10/99)

These four reports all include a generic section on smoking-related safety guidance in a section on all types of fire safety tips. It is recommended that these be replaced with a paragraph incorporating all the recommended messages, i.e.:

> If you smoke, smoke outside. Wherever you smoke, use deep, sturdy ashtrays. Before you throw out butts and ashes, make sure they are out, and dowsing in water or sand is the best way to do that. Check under furniture cushions and in other places people smoke for cigarette butts that may have fallen out of sight. Smoking should not be allowed in a home where oxygen is used. If you smoke, choose fire-safe cigarettes. They are less likely to cause fires. To prevent a deadly cigarette fire, you have to be alert. You won't be if you are sleepy, have been drinking, or have taken medicine or other drugs.

Also, consider whether more targeted messages might make sense, e.g.:

> Your disability/age may mean you will need more time to discover or react to a fire. These recommendations on where, when and how to smoke--and on what to do to keep control of cigarette butts and ashes--are extra important to make sure you are not trapped by a deadly fire.

It is strongly recommended that any such message be developed and reviewed with the full participation and approval of an appropriate advocacy group from the disabled or older-adult community.

The Rural Fire Problem in the United States (FA-180/August 1998)

Because this is a research report rather than a public education piece, it is not a candidate for revised language itself. If or when a public education campaign is developed with a rural focus, then it is recommended that the earlier cited messaging be incorporated in the smoking-related fires section, with the exact version depending on the size of the space available.

The research indicates that smoking does not rank as high among causes of fatal fires in rural areas as in nonrural areas. Heating ranks highest. Nevertheless, smoking is a leading cause of fatal fires in rural areas, and it should receive strong emphasis in a rural-oriented campaign.

Fire Safe Student Housing (FA-228/February 1, 1999)

This is a guide for campus housing administrators and so is not an exact fit for the messaging developed here. Smoking is the subject of two paragraphs on page 11. The research portion is flawed. Falling asleep or passing out is not the primary mechanism for fatal smoking fires. There is no need to identify a problem as primary. The two problems identified--the other is a long smoldering period that hinders detection while people are awake--should have their order reversed, but neither should be identified as primary.

The advice given usefully differentiates between the increasingly common situation of a smoke-free campus --in which indoor smoking is prohibited everywhere and the rules of fire safety essentially consist of rigorously enforcing the ban--and the situation of an unregulated campus, in which case the author's advice should be changed to the recommended all-message paragraph, as cited under the Fire Risks for the Older Adult series.

For more information or copies of this publication, please contact:

Department of Homeland Security
U.S. Fire Administration
16825 South SetonAvenue
Emmitsburg, Maryland 21727
800-561-3356
www.usfa.dhs.gov

FA-302/February 2006

www.ingramcontent.com/pod-product-compliance
Lightning Source LLC
Chambersburg PA
CBHW081222170526

45165CB00009B/2910